Practice

RF/Microwave Engineering

Adel Benleulmi

ISBN 978-1-7773912-0-1

Publisher: Aures Labs Publishing

Cover designer: Mohamed Nedjib Benleulmi

PUBLISHING

CONTENTS

PREFACE

This book is devoted exclusively to exercises and problems in certain areas of RF/microwave engineering. Its aim is to help undergraduate students to prepare for their exams by practicing what they have learned in the classroom. While solutions to the problems presented in this book are provided, it is recommended to make use of study notes to be able to follow along properly. Sections pertaining to some more advanced concepts such as microwave filters would require students to consult specialized textbooks to have access to certain charts and values. Finally, though all solutions presented herein have been previously checked, the presence of errors tends to be ineludible and I would be delighted to receive corrections, suggestions or criticism at author@aureslabs.com.

PART ONE
TRANSMISSION LINES AND WAVEGUIDES

Problem 1. The per-unit-length parameters of two transmission lines, *TL 1* and *TL 2*, are given in the following table:

TL 1	TL 2
$L = 0.3\ \mu H/m$	$L = 0.2\ \mu H/m$
$C = 400\ pF/m$	$C = 200\ pF/m$
$R = 6\ \Omega/m$	$R = 4\ \Omega/m$
$G = 0.02\ S/m$	$G = 0.01\ S/m$

1. Calculate the propagation constant γ and characteristic impedance Z_0 of each line at $f = 700$ MHz.
2. Recalculate these quantities in the absence of loss ($R = G = 0$).

SOLUTION

TL 1:

1. $\gamma = \sqrt{(R + j\omega L)(G + j\omega C)} = 0.38 + j48.18$

   Note: $\gamma = \alpha + j\beta$. Therefore: $\alpha = 0.38$ Np/m and $\beta = 48.18$ rad/m.

   $Z_0 = \sqrt{\dfrac{R + j\omega L}{G + j\omega C}} = 27.39 + j0.09\ \Omega$

2. With $R = G = 0$, $\alpha = 0 \rightarrow \gamma = j\omega\sqrt{LC}$

   $\beta = \omega\sqrt{LC} = 48.18$ rad/m

   $Z_0 = \sqrt{\dfrac{L}{C}} = 27.39\ \Omega$

TL 2:

1. $\gamma = 0.22 + j27.82 \rightarrow \alpha = 0.22$ Np/m and $\beta = 27.82$ rad/m.

   $Z_0 = 31.62 + j0.11\ \Omega$

2. $\beta = 27.82$ rad/m

   $Z_0 = 31.62\ \Omega$

Problem 2. Consider a lossless transmission line of electrical length $\ell = 0.4\lambda$ terminated with a complex load impedance $Z_L = 40 - j20\ \Omega$.

1. With a characteristic impedance $Z_0 = 50\ \Omega$, calculate:
   a) The reflection coefficient at the load.
   b) The SWR on the line.
   c) The input impedance to the line.
   d) The reflection coefficient at the input of the line.
2. Recalculate for $Z_0 = 75\ \Omega$, $\ell = 0.3\lambda$ and $Z_L = 20 + j30\ \Omega$.
3. Find the same quantities using the Smith chart.

SOLUTION

1. With $Z_0 = 50\ \Omega$, $\ell = 0.4\lambda$ and $Z_L = 40 - j20\ \Omega$:

a) The reflection coefficient at the load:

$$\Gamma_L = \frac{Z_L - Z_0}{Z_L + Z_0} = \frac{40 - j\,20 - 50}{40 - j\,20 + 50} = -0.0588 - j0.2353 = 0.243\ \angle 256°$$

b) The SWR on the line:

$$\text{SWR} = \frac{1 + |\Gamma|}{1 - |\Gamma|} = \frac{1 + 0.243}{1 - 0.243} = 1.64$$

c) The input impedance to the line:

$$Z_{in} = Z_0 \frac{Z_L + j\,Z_0\,\tan\beta\ell}{Z_0 + j\,Z_L\,\tan\beta\ell} \quad \text{with } \beta = \frac{2\pi}{\lambda}$$

$$Z_{in} = 50\,\frac{(40 - j\,20) + j\,50\,\tan(2\pi \times 0.4)}{50 + j\,(40 - j\,20)\,\tan(2\pi \times 0.4)} = 72.7 - j19.9\ \Omega$$

Note: For this calculation, your calculator must be set on "RAD".

d) The reflection coefficient at the input of the line:

$$\Gamma_{in} = \frac{Z_{in} - Z_0}{Z_{in} + Z_0} = 0.243\angle 328°$$

Note: To find the angle value in degrees, your calculator must be set on "DEG".

2. For $Z_0 = 75\ \Omega$, $\ell = 0.3\lambda$ and $Z_L = 20 + j30\ \Omega$:

a) $\Gamma_L = 0.629\angle 134°$

b) $\text{SWR} = 4.39$

c) $Z_{in} = 37.1 - j76.4\ \Omega$

d) $\Gamma_{in} = 0.629\angle 278°$

4

3. Using the Smith chart:

a) Γ_L :

Step 1: Normalize $Z_L \rightarrow z_L = Z_L / Z_0$

$$z_L = \frac{40 - j\,20}{50} = 0.8 - j0.4$$

Step 2: Locate and plot z_L on the Smith chart.

Step 3: Extend the line of the normalized impedance z_L.

Step 4: Measure the distance from the center of the chart to z_L then transfer this measurement to the radially scaled parameters (from the center line to the left).

Step 5: From the *RFL. COEFF, E or I* scale, find $|\Gamma_L| = 0.243$. Read the angle of the reflection coefficient from the *ANGLE OF REFLECTION COEFFICIENT IN DEGREES* scale on the perimeter of the chart as $-104°$. Therefore, $\Gamma_L = 0.243\angle 256°$.

b) *SWR* :

From the *SWR* scale, find SWR = 1.64. It is also possible to find the SWR on the horizontal line to the right of the chart's center.

c) Z_{in} :

Step 1: Follow the extended line of the normalized impedance z_L. Since we want to move away from the load, read 0.394λ on the *WAVELENGHTS TOWARD GENERATOR* scale.

Step 2: Add $\ell = 0.4\lambda$ to obtain (by subtracting 0.5λ) 0.294λ on the *WAVELENGHTS TOWARD GENERATOR* scale then draw a radial line from the center of the chart to this value.

Step 3: Draw the constant reflection coefficient magnitude circle and find its intersection with the radial line at $z_{in} = 1.453 - j0.397$ which corresponds approximately to $Z_{in} = 72.7 - j19.9\ \Omega$.

d) Γ_{in} :

Note that $|\Gamma_L| = |\Gamma_{in}| = 0.243$. Read the angle of the reflection coefficient from the *ANGLE OF REFLECTION COEFFICIENT IN DEGREES* scale as $-32°$. Therefore, $\Gamma_{in} = 0.243\angle 328°$.

TL1:

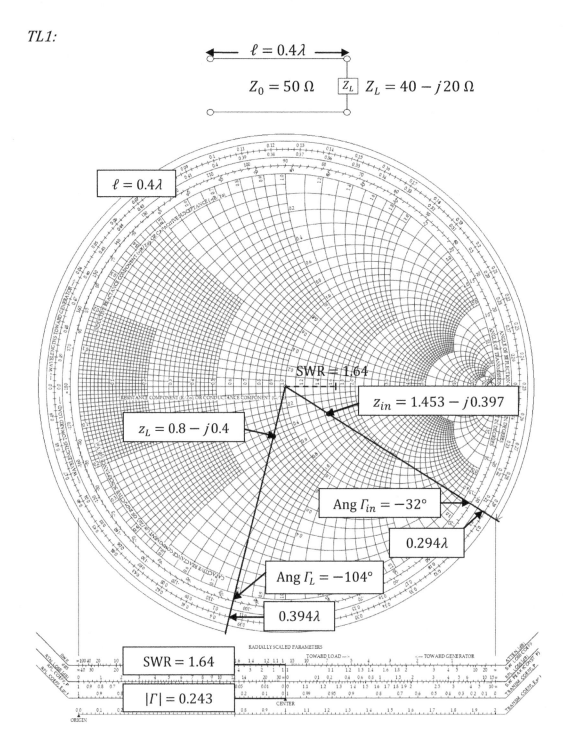

$\ell = 0.4\lambda$

$Z_0 = 50\ \Omega$ $\boxed{Z_L}$ $Z_L = 40 - j20\ \Omega$

a) $\Gamma_L = 0.243\angle 256°$

b) $SWR = 1.64$

c) $Z_{in} = 72.7 - j19.9\ \Omega$

d) $\Gamma_{in} = 0.243\angle 328°$

6

TL2:

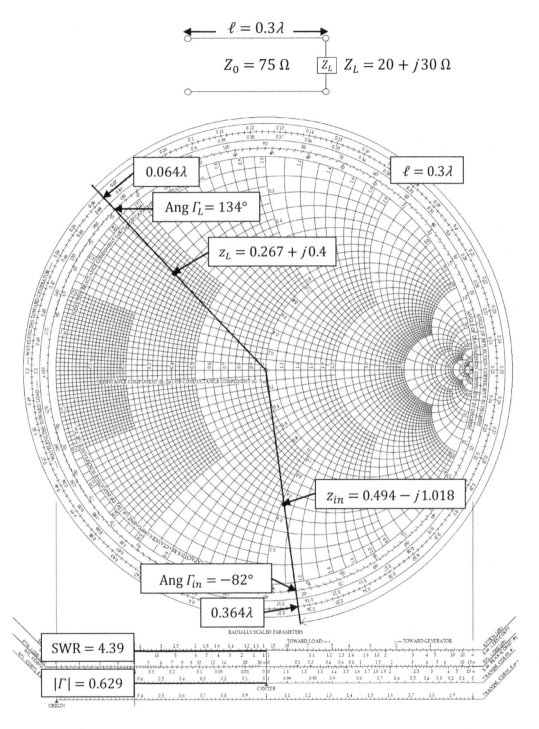

a) $\Gamma_L = 0.629 \angle 134°$

b) SWR = 4.39

c) $Z_{in} = 37.1 - j76.4 \ \Omega$

d) $\Gamma_{in} = 0.629 \angle 278°$

Problem 3. Consider a 4.0 cm coaxial transmission line with a characteristic impedance of 75 Ω terminated with a load impedance of $52.5 + j\,75$ Ω. If the relative permittivity of the line ε_r is 2.56 and the frequency is 2.4 GHz:

1. Calculate:
    a) The input impedance to the line.
    b) The reflection coefficient at the load.
    c) The reflection coefficient at the input.
    d) The SWR on the line.
2. Recalculate for $\ell = 2$ cm and $Z_L = 15 - j\,22.5$ Ω.
3. Find these quantities using the Smith chart.

SOLUTION

1. With $Z_0 = 75$ Ω, $\ell = 4$ cm and $Z_L = 52.5 + j\,75$ Ω:

a) The input impedance to the line:

$$Z_{in} = Z_0 \frac{Z_L + j\,Z_0 \tan\beta\ell}{Z_0 + j\,Z_L \tan\beta\ell}$$

$$\beta = \omega\sqrt{LC} \qquad \text{with} \quad L = \frac{\mu}{2\pi}\ln\frac{b}{a} \text{ and } C = \frac{2\pi\varepsilon}{\ln\frac{b}{a}}$$

$$\beta = \omega\sqrt{\frac{\mu}{2\pi}\ln\frac{b}{a} \times \frac{2\pi\varepsilon}{\ln\frac{b}{a}}} = \omega\sqrt{\mu\varepsilon} = \omega\sqrt{\mu_0\,\varepsilon_0\,\varepsilon_r} = \frac{2\pi f\sqrt{\varepsilon_r}}{c}$$

$$\beta\ell = \frac{2\pi\times2.4\times10^9\times\sqrt{2.56}}{3\times10^8} \times 0.04 = 3.217 \text{ rad}$$

$$Z_{in} = 75\,\frac{(52.5 + j\,75) + j\,75\tan(3.217)}{75 + j\,(52.5 + j\,75)\tan(3.217)} = 61.6 + j\,83.7 \text{ Ω}$$

Note: $\beta\ell$ could also be calculated as follows:

$$\lambda_g = \frac{\lambda_0}{\sqrt{\varepsilon_r}} = \frac{3\times10^8}{2.4\times10^9\times\sqrt{2.56}} = 7.8125 \text{ cm}$$

$$\ell = \frac{4 \text{ cm}}{7.8125 \text{ cm}/\lambda_g} = 0.512\,\lambda_g \rightarrow \beta\ell = \frac{2\pi}{\lambda_g}\times0.512\,\lambda_g = 3.217 \text{ rad}$$

b) The reflection coefficient at the load:

$$\Gamma_L = \frac{Z_L - Z_0}{Z_L + Z_0} = 0.529\,\angle76°$$

c) The reflection coefficient at the input of the line:

$$\Gamma_{in} = \frac{Z_{in} - Z_0}{Z_{in} + Z_0} = 0.529\angle68°$$

d) The SWR on the line:

$$\text{SWR} = \frac{1 + |\Gamma|}{1 - |\Gamma|} = 3.25$$

8

2. For $\ell = 2$ cm and $Z_L = 15 - j22.5\ \Omega$:

   a) $Z_{in} = 138.1 + j183.9\ \Omega$

   b) $\Gamma_L = 0.691\ \angle 215°$

   c) $\Gamma_{in} = 0.691\ \angle 30°$

   d)   SWR $= 5.47$

3. Smith chart solution:

   TL 1:

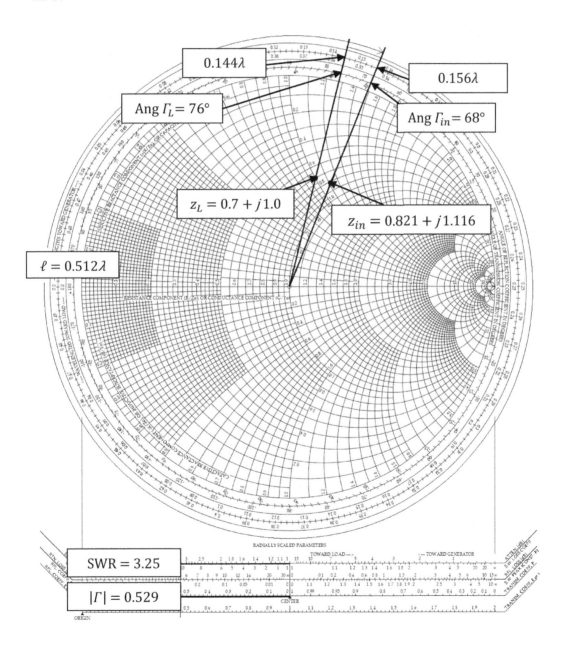

a) $Z_{in} = 61.6 + j83.7 \ \Omega$

b) $\Gamma_L = 0.529 \ \angle 76°$

c) $\Gamma_{in} = 0.529 \ \angle 68°$

d) $SWR = 3.25$

TL 2:

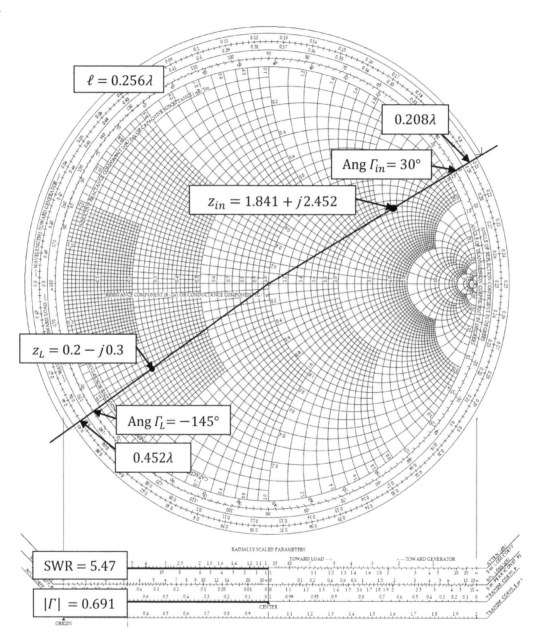

a) $Z_{in} = 138.1 + j183.9 \ \Omega$

b) $\Gamma_L = 0.691 \ \angle 215°$

c) $\Gamma_{in} = 0.691 \ \angle 30°$

d) $SWR = 5.47$

Problem 4. A transmission line has a characteristic impedance of 50 Ω and a reflection coefficient at the load $\Gamma_L = 0.6\angle 122°$. If the input is 0.35λ away from the load:

1. Calculate:
    a) The load impedance.
    b) The reflection coefficient at the input.
    c) The input impedance.

2. Use the Smith chart to find the same quantities.

SOLUTION

1.

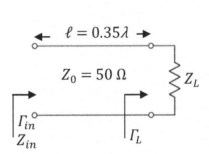

a) $Z_L = Z_0 \dfrac{1+\Gamma_L}{1-\Gamma_L} = 16.0 + j25.5\ \Omega$

b) $\Gamma_{in} = \Gamma_L\, e^{-2j\beta\ell} = 0.6\angle(122 - (2 \times 0.35 \times 360)) = 0.6\angle - 130° = 0.6\angle 230°$

c) $Z_{in} = Z_0 \dfrac{1+\Gamma_{in}}{1-\Gamma_{in}} = 15.0 - j21.6\ \Omega$

2. Smith chart solution:

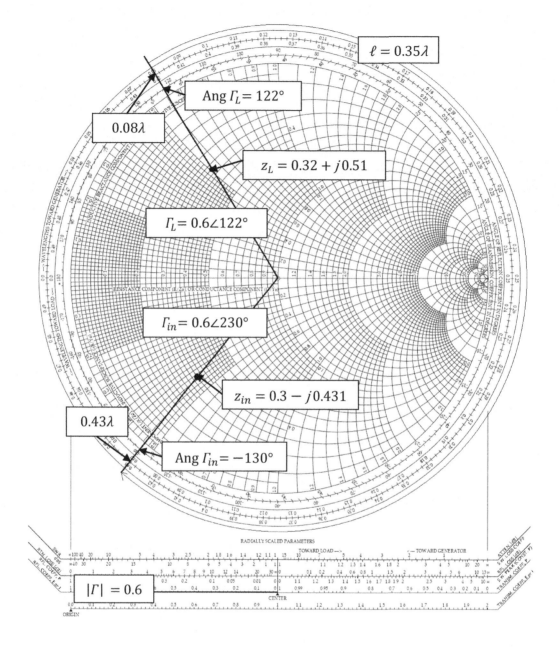

a) $Z_L = 16.0 + j25.5 \ \Omega$

b) $\Gamma_{in} = 0.6 \ \angle 230°$

c) $Z_{in} = 15.0 - j21.6 \ \Omega$

Problem 5. A 50 Ω transmission line of electrical length $\ell = 0.26\lambda$ is terminated with a complex load impedance $Z_L = 20 - j20$ Ω.

1. Find using the Smith chart:
   a) The reflection coefficient at the load.
   b) The load admittance.
   c) The input impedance of the line.
   d) The reflection coefficient at the input of the line.
   e) The distance from the load to the first voltage minimum.
   f) The distance from the load to the first voltage maximum.
   g) The SWR on the line.
2. Find the same quantities for $Z_0 = 75$ Ω, $\ell = 0.38\lambda$ and $Z_L = 22.5 + j37.5$ Ω.

SOLUTION

1.

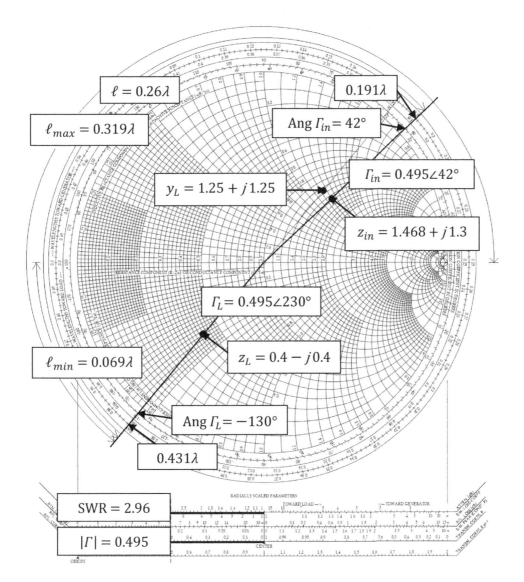

$\ell = 0.26\lambda$

0.191λ

$\ell_{max} = 0.319\lambda$

Ang $\Gamma_{in} = 42°$

$\Gamma_{in} = 0.495\angle 42°$

$y_L = 1.25 + j1.25$

$z_{in} = 1.468 + j1.3$

$\Gamma_L = 0.495\angle 230°$

$\ell_{min} = 0.069\lambda$

$z_L = 0.4 - j0.4$

Ang $\Gamma_L = -130°$

0.431λ

SWR = 2.96

$|\Gamma| = 0.495$

13

a) $\Gamma_L = 0.495 \angle 230°$

b) $Y_L = 25 + j25$ mS

c) $Z_{in} = 73.4 + j65\ \Omega$

d) $\Gamma_{in} = 0.495 \angle 42°$

e) $\ell_{min} = 0.069\lambda$

f) $\ell_{max} = 0.319\lambda$

g) SWR $= 2.96$

2. For $Z_0 = 75\ \Omega$, $\ell = 0.38\lambda$ and $Z_L = 22.5 + j37.5\ \Omega$:

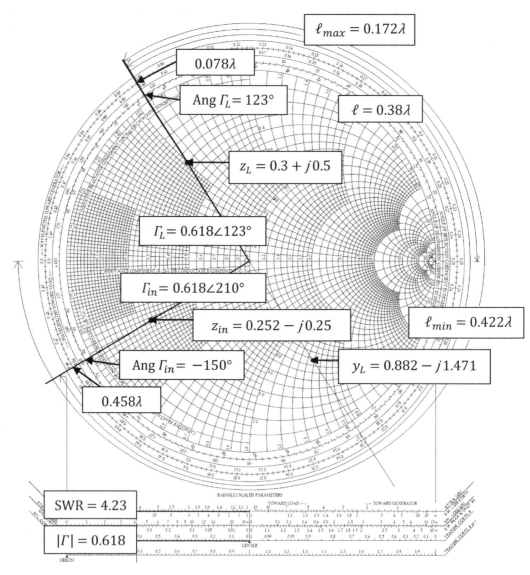

a) $\Gamma_L = 0.618 \angle 123°$

b) $Y_L = 11.8 - j19.6 \text{ mS}$

c) $Z_{in} = 18.9 - j18.8 \, \Omega$

d) $\Gamma_{in} = 0.618 \angle 210°$

e) $\ell_{min} = 0.422\lambda$

f) $\ell_{max} = 0.172\lambda$

g) $\text{SWR} = 4.23$

Note: These results correspond approximately to the analytical calculations.

Problem 6. Using the Smith chart:

1. Find the shortest lengths of a short-circuited 50 Ω line to give the following input impedance:

   a) $Z_{in} = 0$

   b) $Z_{in} = j50\ \Omega$

   c) $Z_{in} = -j20\ \Omega$

   d) $Z_{in} = j35\ \Omega$

   e) $Z_{in} = -j75\ \Omega$

   f) $Z_{in} = j60\ \Omega$

   g) $Z_{in} = \infty$

2. Repeat for an open-circuited length of 75 Ω line.

SOLUTION

1.

   a) $z_{in} = 0 \rightarrow \ell = 0$ or $\ell = 0.5\lambda$

   b) $z_{in} = j1 \rightarrow \ell = 0.125\lambda$

   c) $z_{in} = -j0.4 \rightarrow \ell = 0.439\lambda$

   d) $z_{in} = j0.7 \rightarrow \ell = 0.097\lambda$

   e) $z_{in} = -j1.5 \rightarrow \ell = 0.344\lambda$

   f) $z_{in} = j1.2 \rightarrow \ell = 0.139\lambda$

   g) $z_{in} = \infty \rightarrow \ell = 0.25\lambda$

Note: These results check with $Z_{in} = j\ Z_0\ \tan\beta\ell$.

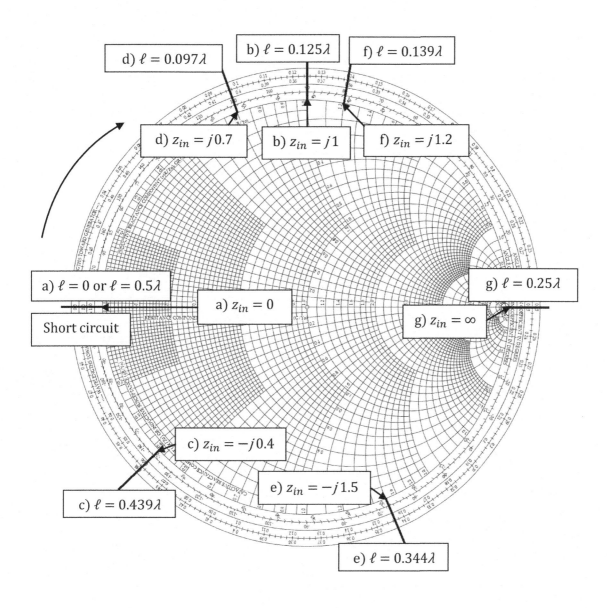

2.

   a) $z_{in} = 0 \rightarrow \ell = 0.25\lambda$

   b) $z_{in} = j0.67 \rightarrow \ell = 0.25\lambda + 0.094\lambda = 0.344\lambda$

   c) $z_{in} = -j0.27 \rightarrow \ell = 0.25\lambda - (0.5 - 0.458)\lambda = 0.208\lambda$

   d) $z_{in} = j0.47 \rightarrow \ell = 0.25\lambda + 0.07\lambda = 0.32\lambda$

   e) $z_{in} = -j1 \rightarrow \ell = 0.25\lambda - (0.5 - 0.375)\lambda = 0.125\lambda$

   f) $z_{in} = j0.8 \rightarrow \ell = 0.25\lambda + 0.107\lambda = 0.357\lambda$

   g) $z_{in} = \infty \rightarrow \ell = 0$ or $\ell = 0.5\lambda$

Note: These results check with $Z_{in} = -j\, Z_0 \cot \beta\ell$.

17

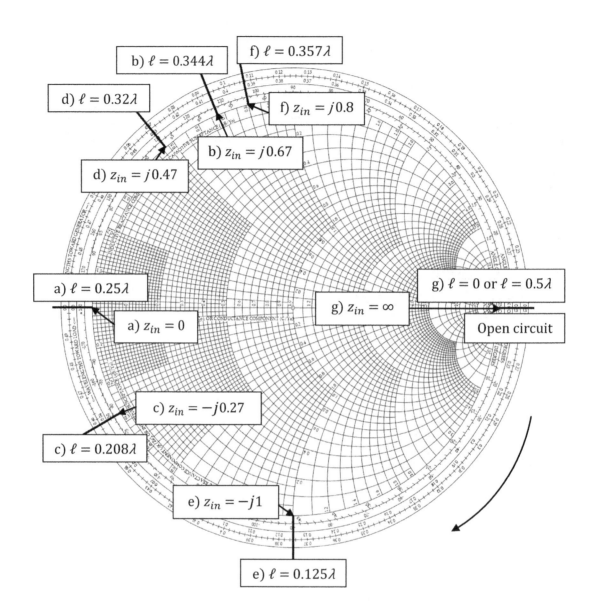

b) $\ell = 0.344\lambda$

f) $\ell = 0.357\lambda$

d) $\ell = 0.32\lambda$

f) $z_{in} = j0.8$

b) $z_{in} = j0.67$

d) $z_{in} = j0.47$

a) $\ell = 0.25\lambda$

g) $\ell = 0$ or $\ell = 0.5\lambda$

a) $z_{in} = 0$

g) $z_{in} = \infty$

Open circuit

c) $z_{in} = -j0.27$

c) $\ell = 0.208\lambda$

e) $z_{in} = -j1$

e) $\ell = 0.125\lambda$

Problem 7. Complete the following table by calculating the SWR, reflection coefficient magnitude and return loss values:

SWR	$|\Gamma|$	RL (dB)
1		
	0.2	
4.0		
		1.94
	1	

SOLUTION

$$RL \text{ (dB)} = -20 \log |\Gamma|$$

$$SWR = \frac{1 + |\Gamma|}{1 - |\Gamma|}$$

$$|\Gamma| = 10^{-RL/20}$$

$$|\Gamma| = \frac{SWR - 1}{SWR + 1}$$

SWR	$|\Gamma|$	RL (dB)
1	0	∞
1.5	0.2	13.98
4.0	0.6	4.44
9.0	0.8	1.94
∞	1	0

Note: You can verify these results using the radially scaled parameters as follows:

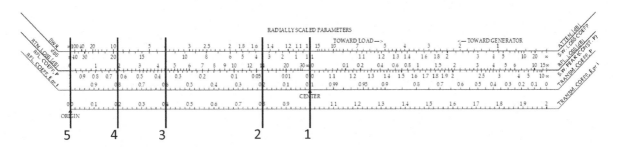

Problem 8. Consider two lossless transmission lines terminated with a 75 Ω and 150 Ω load, respectively.

1. Calculate the two possible values for the characteristic impedance of each line if the SWR on the lines is 4.0.
2. Determine the return loss in dB.

SOLUTION

1. The possible values for the characteristic impedance of each line:

TL 1:

$$Z_0 = ? \quad Z_L = 75 \ \Omega$$

$$|\Gamma| = \frac{SWR - 1}{SWR + 1} = \frac{4 - 1}{4 + 1} = 0.6$$

$$\Gamma = \frac{Z_L - Z_0}{Z_L + Z_0} \quad \rightarrow \quad Z_0 = Z_L \frac{1 - \Gamma}{1 + \Gamma}$$

For $\Gamma = 0.6$:

$$Z_0 = 75 \times \frac{1 - 0.6}{1 + 0.6} = 18.75 \ \Omega$$

For $\Gamma = -0.6$:

$$Z_0 = 75 \times \frac{1 + 0.6}{1 - 0.6} = 300 \ \Omega$$

TL 2:

$$Z_0 = ? \quad Z_L = 150 \ \Omega$$

For $\Gamma = 0.6$:

$$Z_0 = 150 \times \frac{1 - 0.6}{1 + 0.6} = 37.5 \ \Omega$$

For $\Gamma = -0.6$:

$$Z_0 = 150 \times \frac{1 + 0.6}{1 - 0.6} = 600 \ \Omega$$

2. The return loss (dB):

$$RL = -20 \ log \ |\Gamma| = -20 \ log(0.6) = 4.44 \ \text{dB}$$

Problem 9. Find the reflection coefficient magnitude $|\Gamma|$ if the load impedance is purely reactive ($Z_L = jX$) and the characteristic impedance Z_0 is real.

SOLUTION

$$\Gamma = \frac{Z_L - Z_0}{Z_L + Z_0} = \frac{jX - Z_0}{jX + Z_0}$$

$$|\Gamma|^2 = \Gamma\Gamma^* = \frac{(jX - Z_0)(-jX - Z_0)}{(jX + Z_0)(-jX + Z_0)} = \frac{X^2 + Z_0^2}{X^2 + Z_0^2} = 1$$

$$|\Gamma|^2 = 1 \qquad \rightarrow \qquad |\Gamma| = 1$$

This result is shown on the Smith chart in the following figure.

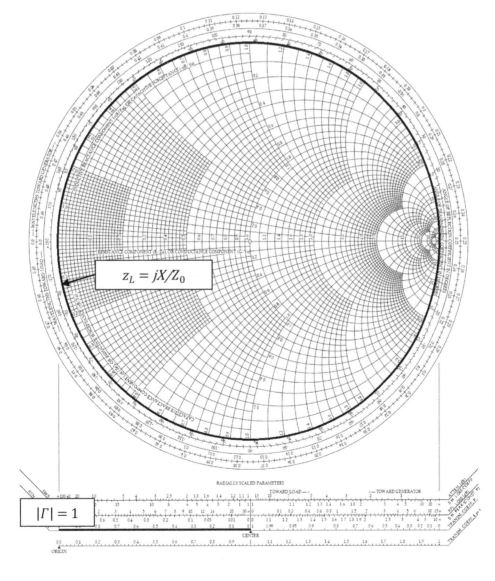

$$z_L = jX/Z_0$$

$$|\Gamma| = 1$$

21

Problem 10. Consider an antenna of an impedance $15 + j\,10\ \Omega$ connected to a $50\ \Omega$ radio transmitter with a $50\ \Omega$ coaxial cable. Knowing that the transmitter can deliver a power of 40 W when connected to a $50\ \Omega$ load:

1. Calculate the power delivered to the antenna.
2. Repeat using the Smith chart for an antenna impedance of $20 - j\,40\ \Omega$.

SOLUTION

1. $\Gamma = \dfrac{Z_L - Z_0}{Z_L + Z_0} = 0.553\ \angle 155°$

$P_{load} = P_{incident} - P_{reflected}$ ⠀⠀⠀⠀⠀⠀ With: $P_{reflected} = P_{incident}\ |\Gamma|^2$

$P_{load} = P_{incident} - P_{incident}\ |\Gamma|^2 = P_{incident}\ (1 - |\Gamma|^2) = 40(1 - (0.553)^2) = 27.8\ \text{W}$

2. Using the Smith chart for an antenna impedance of $20 - j\,40\ \Omega$:

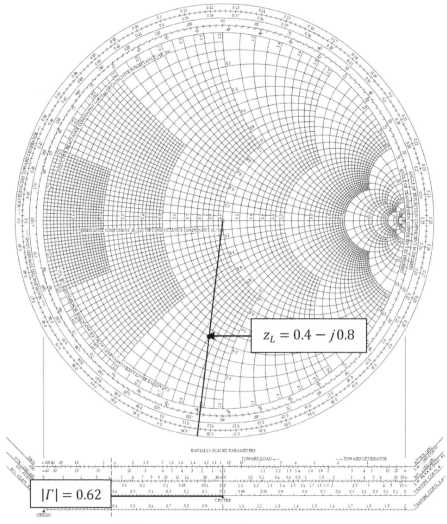

$z_L = 0.4 - j0.8$

$|\Gamma| = 0.62$

$P_{load} = P_{incident}\ (1 - |\Gamma|^2) = 40\ (1 - (0.62)^2) = 24.6\ \text{W}$

Problem 11. Consider the following transmission line:

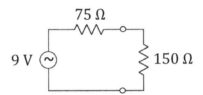

1. Calculate the incident, reflected and transmitted power.
2. Verify that power conservation is satisfied.

SOLUTION

1. The equivalent circuit:

The incident power is:

$$P_{incident} = \frac{1}{2}(75)(\frac{9}{75+75})^2 = 0.135 \text{ W}$$

The reflected power is:

$$P_{reflected} = P_{incident} \ |\Gamma|^2 = 0.135 \left|\frac{150-75}{150+75}\right|^2 = 0.015 \text{ W}$$

The transmitted power is:

$$P_{transmitted} = \frac{1}{2}(150)(\frac{9}{75+150})^2 = 0.12 \text{ W}$$

2. First, calculate the power delivered by the source:

$$P_{source} = \frac{1}{2}\frac{(9)^2}{(75+150)} = 0.18 \text{ W}$$

Then, the power dissipated in the 75 Ω:

$$P_{dissipated} = \frac{1}{2}(75)(\frac{9}{75+150})^2 = 0.06 \text{ W}$$

Power conservation is satisfied since:

$P_{incident} - P_{reflected} = 0.12 \text{ W} = P_{transmitted}$ and

$P_{dissipated} + P_{transmitted} = 0.18 \text{ W} = P_{source}$.

Problem 12. A 10 V source is matched to a 50 Ω transmission line that feeds a load $Z_L = 100$ Ω. If the line is 1.9λ long and has an attenuation constant $\alpha = 0.4$ dB/λ, calculate:

    a) The generator power P_s.

    b) The power lost in the generator P_{R_g}.

    c) The power delivered to the line P_{in}.

    d) The power delivered to the load P_L.

    e) The power lost in the line P_{Loss}.

SOLUTION

a) The generator power $P_s = \frac{1}{2} V_g |I_{in}|$ with $I_{in} = \frac{V_g}{R_g + Z_{in}}$

First, calculate the input impedance Z_{in}:

For a terminated lossy transmission line, $Z_{in} = Z_0 \dfrac{Z_L + Z_0 \tanh \gamma \ell}{Z_0 + Z_L \tanh \gamma \ell}$

$\gamma \ell = (\alpha + j\beta)\, \ell \rightarrow \alpha = 0.4$ dB/$\lambda = 0.4 \times \dfrac{\ln(10)}{20}$ Np/$\lambda = 0.0461$ Np/λ and $\beta \ell = \dfrac{2\pi}{\lambda} \times 1.9\lambda = 324°$.

$\gamma \ell = 0.0875 + j324° \rightarrow Z_{in} = 65.7 - j29.5$ Ω

Then, calculate the input current I_{in}:

$I_{in} = \dfrac{V_g}{R_g + Z_{in}} = 0.0837\angle 14.3°$ A

The generator power $P_s = \frac{1}{2} V_g |I_{in}| = 5\,(0.0837) = 0.42$ W

b) The power lost in R_g :

$P_{R_g} = \frac{1}{2} |I_{in}|^2\, R_g = \frac{1}{2}(0.0837)^2\,(50) = 0.18$ W

c) The power delivered to the line $P_{in} = \dfrac{|V_0^+|^2}{2\,Z_0}\,(1 - |\Gamma(\ell)|^2)\,e^{2\alpha\ell}$

Since the generator is matched to the line: $V_0^+ = \dfrac{V_g}{2}\,e^{-\gamma\ell}$ (phase reference at z = 0)

$|V_0^+| = \left|\dfrac{10}{2}\,e^{-(0.0875+\,j\,324°)}\right| = 4.58\ \text{V}$

$\Gamma(\ell) = \Gamma\,e^{-2\gamma\ell}$ with $\Gamma = \dfrac{Z_L - Z_0}{Z_L + Z_0} = \dfrac{100 - 50}{100 + 50} = 0.333$

$P_{in} = \dfrac{(4.58)^2}{100}\,(e^{2(0.0875)} - (0.333)^2 e^{-2(0.0875)}) = 0.23\ \text{W}$

d) The power delivered to the load P_L:

$P_L = \dfrac{|V_0^+|^2}{2\,Z_0}\,(1 - |\Gamma|^2) = \dfrac{(4.58)^2}{100}\,(1 - (0.333)^2) = 0.19\ \text{W}$

e) The power lost in the line P_{Loss}:

$P_{Loss} = P_{in} - P_L = 0.23 - 0.19 = 0.04\ \text{W}$

Problem 13. The maximum power capacity of a coaxial line is given by:

$$P_{max} = \frac{\pi a^2 E_d^2}{\eta_0} \ln\frac{b}{a}$$

Find the value of b/a that maximizes the maximum power capacity and determine the corresponding characteristic impedance.

SOLUTION

$$P_{max} = \frac{\pi a^2 E_d^2}{\eta_0} \ln\frac{b}{a}$$

$$P_{max} = C\, a^2 \ln\frac{b}{a}$$

$$\frac{dP_{max}}{da} = 2a \ln\frac{b}{a} - \frac{a^2}{a} = 0$$

$$2 \ln\frac{b}{a} - 1 = 0$$

$$\ln\frac{b}{a} = 0.5 \rightarrow \frac{b}{a} = e^{0.5} = 1.65$$

$$Z_0 = \frac{377}{2\pi} \ln\frac{b}{a} = \frac{377\,(0.5)}{2\pi} = 30\ \Omega$$

Problem 14. Consider a section of air-filled R-band waveguide of dimensions $a = 10.922$ cm and $b = 5.461$ cm. The recommended frequency range for this waveguide is 1.70 GHz to 2.60 GHz.
1. Determine the cutoff frequencies of the first two propagating modes.
2. Determine the percentage reduction in bandwidth that the recommended operating range represents relative to the theoretical bandwidth for a single propagating mode.

SOLUTION

1. The cutoff frequencies are given by:

$$f_{c_{mn}} = \frac{c}{2\pi\sqrt{\varepsilon_r}}\sqrt{(\frac{m\pi}{a})^2 + (\frac{n\pi}{b})^2}$$

The lowest order modes $f_{c_{10}}$ and $f_{c_{20}}$ are:

$$f_{c_{10}} = \frac{c}{2a} = 1.37 \text{ GHz}$$

$$f_{c_{20}} = \frac{c}{a} = 2.75 \text{ GHz}$$

2. The fractional bandwidth from $f_{c_{10}}$ to $f_{c_{20}}$ is: $\frac{2.75 - 1.37}{(2.75 + 1.37)/2} = 67\%$

The fractional bandwidth of the recommended operating range is: $\frac{2.6 - 1.7}{(2.6 + 1.7)/2} = 42\%$

The reduction in bandwidth is then 25%.

Problem 15. Consider a 50 Ω copper stripline transmission line with a ground plane separation b = 1.524 mm and a relative dielectric constant ε_r = 2.2. If the frequency f = 4.7 GHz, dielectric loss tangent $\tan \delta$ = 0.001 and conductor thickness t = 1 μm:

1. Determine the width W of this transmission line.
2. Find the guide wavelength on the line and the total attenuation in dB/λ.

SOLUTION

1. First, calculate $\sqrt{\varepsilon_r}\, Z_0$:

$$\sqrt{\varepsilon_r}\, Z_0 = \sqrt{2.2} \times 50 = 74.2 < 120, \text{ then:}$$

$$\frac{W}{b} = \frac{30\pi}{\sqrt{\varepsilon_r}\, Z_0} - 0.441$$

$$\frac{W}{b} = \frac{30\pi}{74.2} - 0.441 = 0.83 \rightarrow W = 0.126 \text{ cm}$$

2. The guide wavelength on the line λ_g:

$$\lambda_g = \frac{\lambda_0}{\sqrt{\varepsilon_r}} = \frac{c}{f\sqrt{\varepsilon_r}} = \frac{3 \times 10^8}{(4.7 \times 10^9)\sqrt{2.2}} = 4.30 \text{ cm}$$

The total attenuation in dB/λ:

At 4.7 GHz, the wave number k is:

$$k = \frac{2\pi f \sqrt{\varepsilon_r}}{c} = \frac{2\pi(4.7 \times 10^9)\sqrt{2.2}}{3 \times 10^8} = 146 \text{ m}^{-1}$$

The dielectric attenuation α_d is:

$$\alpha_d = \frac{k \tan \delta}{2} = \frac{(146)(0.001)}{2} = 0.073 \text{ Np/m}$$

The surface resistance of copper R_S at 4.7 GHz is:

$$R_S = \sqrt{\frac{\omega \mu}{2\sigma}} \text{ with the vacuum permeability } \mu_0 = 4\pi \times 10^{-7} \text{ H/m and the conductivity of}$$

copper $\sigma = 5.813 \times 10^7$ S/m.

$$R_S = \sqrt{\frac{2\pi(4.7 \times 10^9)(4\pi \times 10^{-7})}{2(5.813 \times 10^7)}} = 0.018 \ \Omega$$

The conductor attenuation α_C is:

$$\alpha_c = \frac{2.7 \times 10^{-3} \, R_s \, \varepsilon_r \, Z_0}{30\pi(b-t)} A \qquad \text{with} \quad A = 1 + \frac{2W}{b-t} + \frac{1}{\pi}\frac{b+t}{b-t} \, ln(\frac{2b-t}{t}) = 5.21$$

$$\alpha_c = \frac{2.7 \times 10^{-3} \times 0.018 \times 2.2 \times 50 \times 5.21}{30\pi(1.524\times10^{-3}-10^{-6})} = 0.194 \text{ Np/m}$$

The total attenuation constant α is:

$$\alpha = \alpha_d + \alpha_c = 0.073 + 0.194 = 0.267 \text{ Np/m}$$

In dB:

$$\alpha(\text{dB}) = 20 \, log \, e^{\alpha} = 2.32 \text{ dB/m}$$

The attenuation in dB/λ is:

$$\alpha(\text{dB}) = 2.32 \times 0.043 = 0.1 \text{ dB/}\lambda$$

Problem 16. Consider a 50 Ω copper microstrip transmission line designed on a substrate with the following parameters: $\varepsilon_r = 6.15$, $tan\,\delta = 0.002$ and $h = 1.28$ mm. With the operating frequency $f = 10$ GHz:

1. Find the width W of this transmission line.
2. Determine the guide wavelength on the line and the total attenuation in dB/m.

SOLUTION

1. There are two possibilities:

For $\dfrac{W}{d} < 2$ $\rightarrow$ $\dfrac{W}{d} = \dfrac{8e^A}{e^{2A} - 2}$

For $\dfrac{W}{d} \geq 2$ $\rightarrow$ $\dfrac{W}{d} = \dfrac{2}{\pi}\left[B - 1 - ln(2B - 1) + \dfrac{\varepsilon_r - 1}{2\varepsilon_r}\left\{ln(B - 1) + 0.39 - \dfrac{0.61}{\varepsilon_r}\right\}\right]$

With:

$A = \dfrac{Z_0}{60}\sqrt{\dfrac{\varepsilon_r + 1}{2}} + \dfrac{\varepsilon_r - 1}{\varepsilon_r + 1}\left(0.23 + \dfrac{0.11}{\varepsilon_r}\right)$

$B = \dfrac{377\,\pi}{2Z_0\sqrt{\varepsilon_r}}$

First, try $\dfrac{W}{d} < 2$:

$A = \dfrac{50}{60}\sqrt{\dfrac{6.15 + 1}{2}} + \dfrac{6.15 - 1}{6.15 + 1}\left(0.23 + \dfrac{0.11}{6.15}\right) = 1.754$ $\rightarrow$ $\dfrac{W}{d} = \dfrac{8e^{1.754}}{e^{2(1.754)} - 2} = 1.47$

The condition that $\dfrac{W}{d} < 2$ is satisfied. Then:

$W = 1.47 \times 1.28 = 1.88$ mm

2. The guide wavelength on the line λ_g:

$\lambda_g = \dfrac{c}{f\sqrt{\varepsilon_{eff}}}$ with $\varepsilon_{eff} = \dfrac{\varepsilon_r + 1}{2} + \dfrac{\varepsilon_r - 1}{2}\dfrac{1}{\sqrt{1 + 12\frac{d}{W}}}$

$\varepsilon_{eff} = 4.43$

$\lambda_g = 1.43$ cm

The total attenuation constant $\alpha = \alpha_d + \alpha_c$

$$\alpha_d = \frac{k_0 \varepsilon_r \left(\varepsilon_{eff} - 1\right) \tan \delta}{2\sqrt{\varepsilon_{eff}} \left(\varepsilon_r - 1\right)} \qquad \text{with} \quad k_0 = \frac{2\pi f}{c} = 209.4 \text{ m}^{-1}$$

$\alpha_d = 0.407$ Np/m

$$\alpha_c = \frac{R_S}{Z_0 W} \qquad \text{with} \quad R_S = \sqrt{\frac{\omega \mu}{2\sigma}} = \sqrt{\frac{2\pi(10\times10^9)(4\pi\times10^{-7})}{2(5.813\times10^7)}} = 0.026 \ \Omega$$

$\alpha_c = 0.277$ Np/m

$\alpha = \alpha_d + \alpha_c = 0.407 + 0.277 = 0.684$ Np/m

The total attenuation in dB:

$\alpha(\text{dB}) = 20 \ log \ e^{\alpha} = 5.94$ dB/m

Problem 17. A 50 Ω printed transmission line that is 13λ long is to be used in a microwave antenna feed network operating at 6.4 GHz. Two possible choices are given in the following table:

Copper microstrip	Copper stripline
$d = 0.25$ cm	$b = 0.50$ cm
$\varepsilon_r = 3.0$	$\varepsilon_r = 3.0$
$\tan \delta = 0.001$	$\tan \delta = 0.001$
$t = 17$ μm	$t = 17$ μm

Determine which line should be used if total attenuation is to be minimized.

SOLUTION

1. The microstrip TL:

Try $\quad \dfrac{W}{d} > 2 \qquad \rightarrow \qquad B = \dfrac{377\,\pi}{2Z_0\sqrt{\varepsilon_r}} = 6.84$

$$\frac{W}{d} = \frac{2}{\pi}\left[B - 1 - ln(2B - 1) + \frac{\varepsilon_r - 1}{2\varepsilon_r}\left\{ln(B - 1) + 0.39 - \frac{0.61}{\varepsilon_r}\right\}\right] = 2.51 > 2$$

$W = 2.51 \times 0.25 = 0.628$ cm

$$\varepsilon_{eff} = \frac{\varepsilon_r + 1}{2} + \frac{\varepsilon_r - 1}{2}\frac{1}{\sqrt{1 + 12\frac{d}{W}}} = 2.42 \qquad \rightarrow \qquad \lambda_g = \frac{c}{f\sqrt{\varepsilon_{eff}}} = 3.02 \text{ cm}$$

The total attenuation constant: $\alpha = \alpha_d + \alpha_c$

$$\alpha_d = \frac{k_0\varepsilon_r\,(\varepsilon_{eff} - 1)\tan\delta}{2\sqrt{\varepsilon_{eff}}\,(\varepsilon_r - 1)} \qquad \text{with} \quad k_0 = \frac{2\pi f}{c} = 134.04 \text{ m}^{-1}$$

$\alpha_d = 0.092$ Np/m

$$\alpha_c = \frac{R_S}{Z_0 W} \qquad \text{with} \quad R_S = \sqrt{\frac{\omega\mu_0}{2\sigma}} = 0.0208 \text{ Ω}$$

$\alpha_c = 0.066$ Np/m

$\alpha = 0.092 + 0.066 = 0.158$ Np/m

The total microstrip loss:

Loss $= (0.158)$ [Np/m] $\times (13\lambda_g) \times (0.0302)$ [m/λ_g] $\times (8.686)$ [dB/Np] $= 0.54$ dB

2. The stripline TL:

$$\lambda_g = \frac{c}{f\sqrt{\varepsilon_r}} = \frac{3 \times 10^8}{(6.4 \times 10^9)\sqrt{3}} = 2.71 \text{ cm}$$

$$\sqrt{\varepsilon_r} \, Z_0 = 86.6 < 120, \text{ then:}$$

$$\frac{W}{b} = \frac{30\pi}{\sqrt{\varepsilon_r} \, Z_0} - 0.441 = 0.647 \rightarrow W = 0.324 \text{ cm}$$

The total attenuation constant: $\alpha = \alpha_d + \alpha_c$

$$\alpha_d = \frac{k \tan \delta}{2} \quad \text{with} \quad k = \sqrt{\varepsilon_r} \, k_0$$

$$\alpha_d = 0.116 \text{ Np/m}$$

$$\alpha_c = \frac{2.7 \times 10^{-3} \, R_s \, \varepsilon_r \, Z_0 \, A}{30\pi(b-t)} \qquad \text{with} \quad A = 1 + \frac{2W}{b-t} + \frac{1}{\pi} \frac{b+t}{b-t} \ln(\frac{2b-t}{t}) = 4.34$$

$$\alpha_c = 0.078 \text{ Np/m}$$

$$\alpha = \alpha_d + \alpha_c = 0.116 + 0.078 = 0.194 \text{ Np/m}$$

The total stripline loss:

$$\text{Loss} = (0.194) \, [\text{Np/m}] \times (13\lambda_g) \times (0.0271) \, [\text{m}/\lambda_g] \times (8.686) \, [\text{dB/Np}] = 0.59 \text{ dB}$$

Therefore, the microstrip line should be used.

PART TWO
MICROWAVE FILTERS

Problem 1. Design composite low-pass filters by the image parameter method with the following specifications:

Filter 1	Filter 2
$R_0 = 50 \ \Omega$	$R_0 = 75 \ \Omega$
$f_c = 70 \ \text{MHz}$	$f_c = 23 \ \text{MHz}$
$f_\infty = 72 \ \text{MHz}$	$f_\infty = 23.8 \ \text{MHz}$

SOLUTION

Filter 1:

For the constant-*k* section:

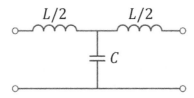

$$L = \frac{2R_0}{\omega_c} = 227.36 \ \text{nH}$$

$$C = \frac{2}{\omega_c R_0} = 90.946 \ \text{pF}$$

For the *m*-derived sharp-cutoff section:

$$m = \sqrt{1 - \left(\frac{f_c}{f_\infty}\right)^2} = 0.2341$$

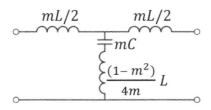

$$mL/2 = 26.6 \ \text{nH}$$

$$mC = 21.3 \ \text{pF}$$

$$\frac{(1-m^2)}{4m} L = 229.5 \ \text{nH}$$

For the $m = 0.6$ matching sections:

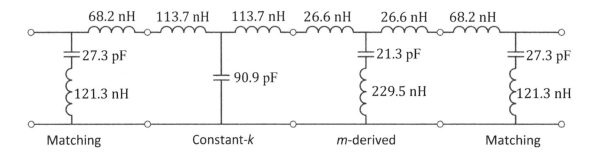

$mL/2 = 68.2 \text{ nH}$

$mC/2 = 27.3 \text{ pF}$

$\frac{(1-m^2)}{2m} L = 121.3 \text{ nH}$

The completed filter circuit is:

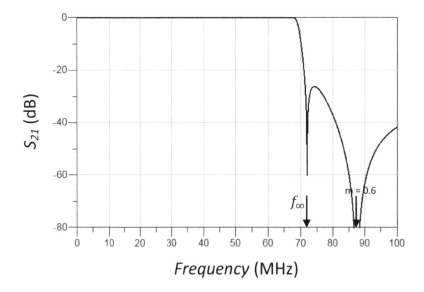

Note: The frequency response of this filter is shown in the following figure.

Filter 2:

For the constant-*k* section:

$$L = \frac{2R_0}{\omega_c} = 1.038 \ \mu H$$

$$C = \frac{2}{\omega_c R_0} = 184.527 \ pF$$

For the *m*-derived sharp-cutoff section:

$$m = \sqrt{1 - (\frac{f_c}{f_\infty})^2} = 0.2571$$

$$mL/2 = 133.4 \ nH$$

$$mC = 47.4 \ pF$$

$$\frac{(1 - m^2)}{4m} L = 0.94 \ \mu H$$

For the *m* = 0.6 matching sections:

$$mL/2 = 311.4 \ nH$$

$$mC/2 = 55.4 \ pF$$

$$\frac{(1 - m^2)}{2m} L = 553.6 \ nH$$

The completed filter circuit is:

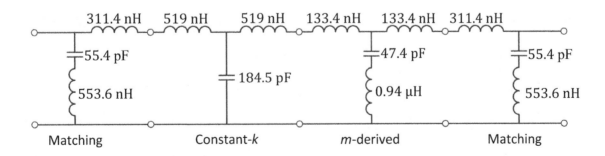

Note: The frequency response of this filter is shown in the following figure.

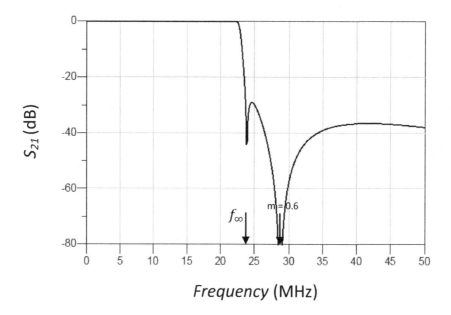

Problem 2. Design composite high-pass filters by the image parameter method with the following specifications:

Filter 1	Filter 2
$R_0 = 75 \, \Omega$	$R_0 = 50 \, \Omega$
$f_c = 62 \text{ MHz}$	$f_c = 23 \text{ MHz}$
$f_\infty = 60 \text{ MHz}$	$f_\infty = 22.4 \text{ MHz}$

SOLUTION

Filter 1:

For the constant-k section:

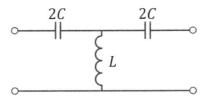

$$L = \frac{R_0}{2\omega_c} = 96.26 \text{ nH}$$

$$C = \frac{1}{2\omega_c R_0} = 17.113 \text{ pF}$$

For the m-derived sharp-cutoff section:

$$m = \sqrt{1 - \left(\frac{f_\infty}{f_c}\right)^2} = 0.2519$$

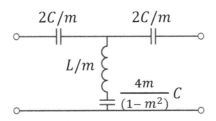

$$2C/m = 135.9 \text{ pF}$$

$$L/m = 382.1 \text{ nH}$$

$$\frac{4m}{(1-m^2)} C = 18.4 \text{ pF}$$

For the $m = 0.6$ matching sections:

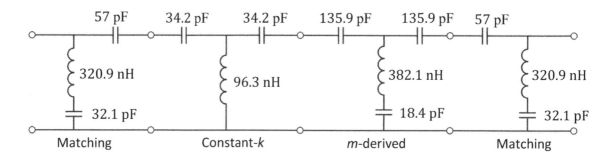

$2C/m = 57.0$ pF

$2L/m = 320.9$ nH

$\dfrac{2m}{(1-m^2)}\,C = 32.1$ pF

The completed filter circuit is:

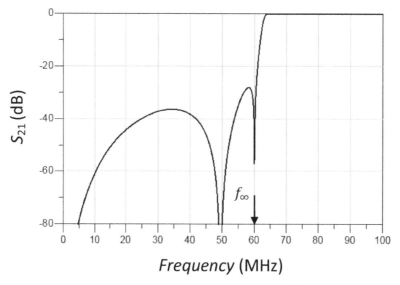

Note: The frequency response of this filter is shown in the following figure.

Filter 2:

For the constant-*k* section:

$$L = \frac{R_0}{2\omega_c} = 172.99 \text{ nH}$$

$$C = \frac{1}{2\omega_c R_0} = 69.198 \text{ pF}$$

For the *m*-derived sharp-cutoff section:

$$m = \sqrt{1 - (\frac{f_\infty}{f_c})^2} = 0.2269$$

$$2C/m = 609.9 \text{ pF}$$

$$L/m = 762.4 \text{ nH}$$

$$\frac{4m}{(1-m^2)}C = 66.2 \text{ pF}$$

For the *m* = 0.6 matching sections:

$$2C/m = 230.7 \text{ pF}$$

$$2L/m = 576.6 \text{ nH}$$

$$\frac{2m}{(1-m^2)}C = 129.7 \text{ pF}$$

The completed filter circuit is:

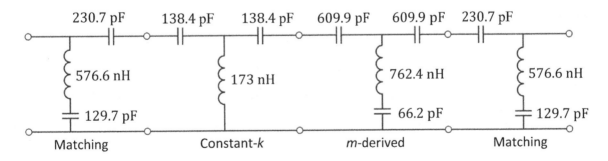

43

Note: The frequency response of this filter is shown in the following figure.

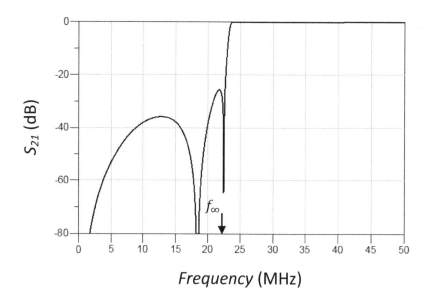

Problem 3. Design low-pass maximally flat lumped-element filters with the following specifications:

	Filter 1	Filter 2
Passband	0 to 3.2 GHz	0 to 1.36 GHz
Attenuation	$\alpha > 20$ dB at 6.4 GHz	$\alpha > 20$ dB at 2.3 GHz
Characteristic impedance	$Z_0 = 75\ \Omega$	$Z_0 = 50\ \Omega$

SOLUTION

Filter 1:

$$\left|\frac{\omega}{\omega_c}\right| - 1 = \frac{6.4}{3.2} - 1 = 1$$

- From the "Attenuation versus normalized frequency for maximally flat filter prototypes" figure, find $N = 4$ for $\alpha > 20$ dB.

Note: The filter's order can be calculated as follows:

$$N = \frac{log\left(10^{A/10}-1\right)}{2\,log\left(\frac{\omega}{\omega_c}\right)} = \frac{log\left(10^{20/10}-1\right)}{2\,log\left(\frac{6.4}{3.2}\right)} = 3.31 \rightarrow N = 4$$

- From the "Element values for maximally flat low-pass filter prototypes" table, find:

$g_1 = 0.7654$

$g_2 = 1.8478$

$g_3 = 1.8478$

$g_4 = 0.7654$

Note: The element values can be calculated with $g_k = 2\sin(\frac{(2k-1)\pi}{2N})$.

The low-pass filter:

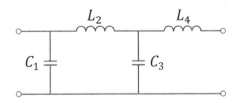

$$C_1 = \frac{g_1}{R_0 \omega_c} = 0.508 \text{ pF}$$

$$L_2 = \frac{R_0 g_2}{\omega_c} = 6.893 \text{ nH}$$

$$C_3 = \frac{g_3}{R_0 \omega_c} = 1.225 \text{ pF}$$

$$L_4 = \frac{R_0 g_4}{\omega_c} = 2.855 \text{ nH}$$

Note: The frequency response of this filter is shown in the following figure.

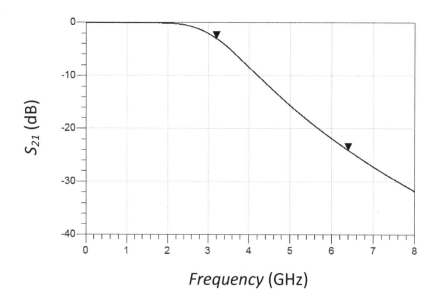

Filter 2:

$$\left| \frac{\omega}{\omega_c} \right| - 1 = 0.691$$

- $N = 5$ for $\alpha > 20$ dB.

$$g_1 = 0.618$$

$$g_2 = 1.618$$

$$g_3 = 2$$

$$g_4 = 1.618$$

$$g_5 = 0.618$$

The low-pass filter:

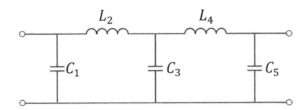

$$C_1 = \frac{g_1}{R_0 \omega_c} = 1.446 \text{ pF}$$

$$L_2 = \frac{R_0 g_2}{\omega_c} = 9.467 \text{ nH}$$

$$C_3 = \frac{g_3}{R_0 \omega_c} = 4.681 \text{ pF}$$

$$L_4 = \frac{R_0 g_4}{\omega_c} = 9.467 \text{ nH}$$

$$C_5 = \frac{g_5}{R_0 \omega_c} = 1.446 \text{ pF}$$

Note: The frequency response of this filter is shown in the following figure.

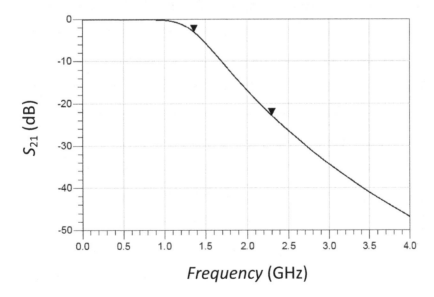

Problem 4. Design high-pass maximally flat lumped-element filters with the following specifications:

	Filter 1	Filter 2
Cutoff frequency	6.4 GHz	3.9 GHz
Attenuation	$\alpha > 20$ dB at 3.76 GHz	$\alpha > 20$ dB at 1.4 GHz
Characteristic impedance	$Z_0 = 50\ \Omega$	$Z_0 = 75\ \Omega$

SOLUTION

Low-pass to high-pass transformation:

Filter 1:

$$\omega \leftarrow -\frac{\omega_c}{\omega}$$

$$\left|\frac{\omega}{\omega_c}\right| - 1 = \frac{6.4}{3.76} - 1 = 0.702$$

- From the "Attenuation versus normalized frequency for maximally flat filter prototypes" figure, find $N = 5$ for $\alpha > 20$ dB.

- From the "Element values for maximally flat low-pass filter prototypes" table, find:

$g_1 = 0.618$

$g_2 = 1.618$

$g_3 = 2$

$g_4 = 1.618$

$g_5 = 0.618$

The high-pass filter:

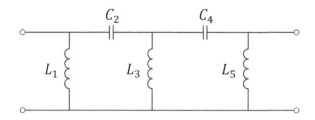

$$L_1 = \frac{R_0}{g_1\omega_c} = 2.012 \text{ nH}$$

$$C_2 = \frac{1}{g_2 R_0 \omega_c} = 0.307 \text{ pF}$$

$$L_3 = \frac{R_0}{g_3\omega_c} = 0.622 \text{ nH}$$

$$C_4 = \frac{1}{g_4 R_0 \omega_c} = 0.307 \text{ pF}$$

$$L_5 = \frac{R_0}{g_5\omega_c} = 2.012 \text{ nH}$$

Note: The frequency response of this filter is shown in the following figure.

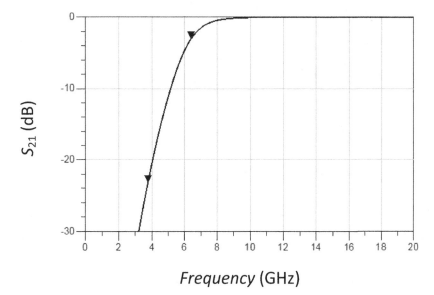

Filter 2:

$$\left|\frac{\omega}{\omega_c}\right| - 1 = 1.786$$

- $N = 3$ for $\alpha > 20$ dB.

$$g_1 = 1$$

$$g_2 = 2$$

$$g_3 = 1$$

The high-pass filter:

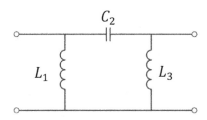

$$L_1 = \frac{R_0}{g_1 \omega_c} = 3.061 \text{ nH}$$

$$C_2 = \frac{1}{g_2 R_0 \omega_c} = 0.272 \text{ pF}$$

$$L_3 = \frac{R_0}{g_3 \omega_c} = 3.061 \text{ nH}$$

Note: The frequency response of this filter is shown in the following figure.

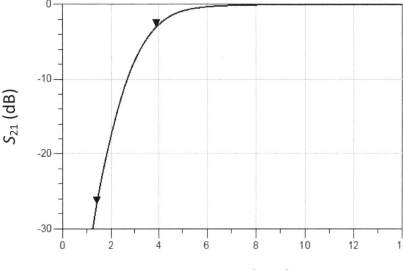

> **Problem 5.** Design bandpass maximally flat lumped-element filters with the following specifications:
>
> | | Filter 1 | Filter 2 |
> | --- | --- | --- |
> | Order | $N = 3$ | $N = 5$ |
> | Center frequency | $f_0 = 3.2$ GHz | $f_0 = 2$ GHz |
> | Bandwidth | 400 MHz | 100 MHz |
> | Impedance | $Z_0 = 50\ \Omega$ | $Z_0 = 75\ \Omega$ |

SOLUTION

Low-pass to bandpass transformation:

Filter 1:

$$\Delta = \frac{\omega_2 - \omega_1}{\omega_0} = \frac{400 \times 10^6}{3.2 \times 10^9} = 0.125$$

- From the "Element values for maximally flat low-pass filter prototypes" table, find:

$g_1 = 1$

$g_2 = 2$

$g_3 = 1$

The bandpass filter:

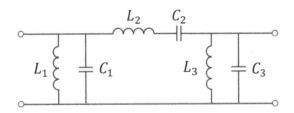

$$L_1 = \frac{\Delta Z_0}{\omega_0 g_1} = 0.311 \text{ nH}$$

$$C_1 = \frac{g_1}{\omega_0 \Delta Z_0} = 7.958 \text{ pF}$$

$$L_2 = \frac{g_2 Z_0}{\omega_0 \Delta} = 39.789 \text{ nH}$$

$$C_2 = \frac{\Delta}{\omega_0 g_2 Z_0} = 0.062 \text{ pF}$$

$$L_3 = \frac{\Delta Z_0}{\omega_0 g_3} = 0.311 \text{ nH}$$

$$C_3 = \frac{g_3}{\omega_0 \Delta Z_0} = 7.958 \text{ pF}$$

Note: The frequency response of this filter is shown in the following figure.

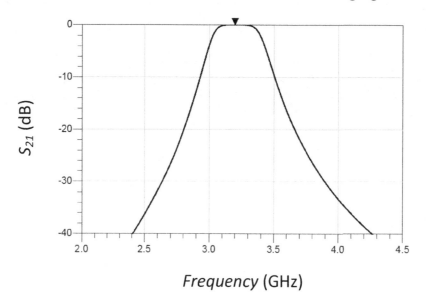

Frequency (GHz)

Filter 2:

$$\Delta = \frac{\omega_2 - \omega_1}{\omega_0} = 0.05$$

- From the "Element values for maximally flat low-pass filter prototypes" table, find:

$g_1 = 0.618$

$g_2 = 1.618$

$g_3 = 2$

$g_4 = 1.618$

$g_5 = 0.618$

The bandpass filter:

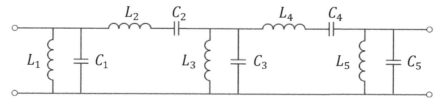

$$L_1 = \frac{\Delta Z_0}{\omega_0 g_1} = 0.483 \text{ nH}$$

$$C_1 = \frac{g_1}{\omega_0 \Delta Z_0} = 13.114 \text{ pF}$$

$$L_2 = \frac{g_2 Z_0}{\omega_0 \Delta} = 193.135 \text{ nH}$$

$$C_2 = \frac{\Delta}{\omega_0 g_2 Z_0} = 0.0328 \text{ pF}$$

$$L_3 = \frac{\Delta Z_0}{\omega_0 g_3} = 0.149 \text{ nH}$$

$$C_3 = \frac{g_3}{\omega_0 \Delta Z_0} = 42.441 \text{ pF}$$

$$L_4 = \frac{g_4 Z_0}{\omega_0 \Delta} = 193.135 \text{ nH}$$

$$C_4 = \frac{\Delta}{\omega_0 g_4 Z_0} = 0.0328 \text{ pF}$$

$$L_5 = \frac{\Delta Z_0}{\omega_0 g_5} = 0.483 \text{ nH}$$

$$C_5 = \frac{g_5}{\omega_0 \Delta Z_0} = 13.114 \text{ pF}$$

Note: The frequency response of this filter is shown in the following figure.

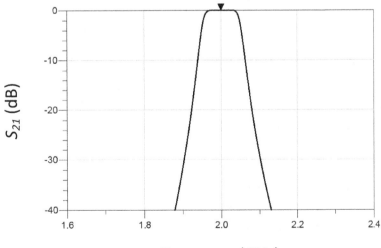

Problem 6. Design high-pass lumped-element filters with the following specifications:

	Filter 1 : 3 dB equal-ripple	Filter 2 : 0.5 dB equal-ripple
Cutoff frequency	$f_c = 2.5$ GHz	$f_c = 6.35$ GHz
Attenuation	$\alpha > 30$ dB at 1.66 GHz	$\alpha > 20$ dB at 5.24 GHz
Characteristic impedance	$Z_0 = 50\ \Omega$	$Z_0 = 75\ \Omega$

SOLUTION

Low-pass to high-pass transformation:

Filter 1:

$$\omega \leftarrow -\frac{\omega_c}{\omega}$$

$$\left|\frac{\omega}{\omega_c}\right| - 1 = \frac{2.5}{1.66} - 1 = 0.506$$

- From the "Attenuation versus normalized frequency for equal-ripple filter prototypes (3 dB ripple level)" figure, find $N = 5$ for $\alpha > 30$ dB.

- From the "Element values for equal-ripple low-pass filter prototypes (3 dB ripple)" table, find:

$g_1 = 3.4817$

$g_2 = 0.7618$

$g_3 = 4.5381$

$g_4 = 0.7618$

$g_5 = 3.4817$

The high-pass filter:

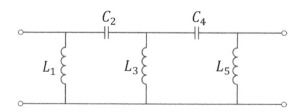

$$L_1 = \frac{R_0}{\omega_c g_1} = 0.914 \text{ nH}$$

$$C_2 = \frac{1}{R_0 \omega_c g_2} = 1.671 \text{ pF}$$

$$L_3 = \frac{R_0}{\omega_c g_3} = 0.701 \text{ nH}$$

$$C_4 = \frac{1}{R_0 \omega_c g_4} = 1.671 \text{ pF}$$

$$L_5 = \frac{R_0}{\omega_c g_5} = 0.914 \text{ nH}$$

Note: The frequency response of this filter is shown in the following figure.

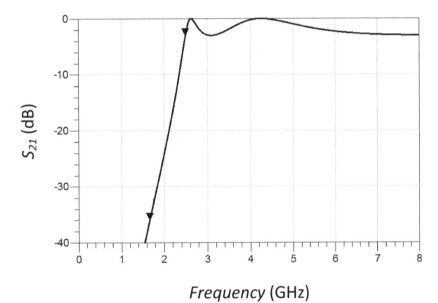

Filter 2:

$$\omega \leftarrow -\frac{\omega_c}{\omega}$$

$$\left|\frac{\omega}{\omega_c}\right| - 1 = \frac{6.35}{5.24} - 1 = 0.212$$

- From the "Attenuation versus normalized frequency for equal-ripple filter prototypes (0.5 dB ripple level)" figure, find $N = 7$ for $\alpha > 20$ dB.

- From the "Element values for equal-ripple low-pass filter prototypes (0.5 dB ripple)" table, find:

$g_1 = 1.7372$

$g_2 = 1.2583$

$g_3 = 2.6381$

$g_4 = 1.3444$

$g_5 = 2.6381$

$g_6 = 1.2583$

$g_7 = 1.7372$

The high-pass filter:

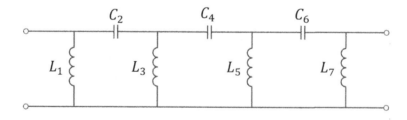

$$L_1 = \frac{R_0}{\omega_c g_1} = 1.082 \text{ nH}$$

$$C_2 = \frac{1}{R_0 \omega_c g_2} = 0.266 \text{ pF}$$

$$L_3 = \frac{R_0}{\omega_c g_3} = 0.713 \text{ nH}$$

$$C_4 = \frac{1}{R_0 \omega_c g_4} = 0.249 \text{ pF}$$

$$L_5 = \frac{R_0}{\omega_c g_5} = 0.713 \text{ nH}$$

$$C_6 = \frac{1}{R_0 \omega_c g_6} = 0.266 \text{ pF}$$

$$L_7 = \frac{R_0}{\omega_c g_7} = 1.082 \text{ nH}$$

Note: The frequency response of this filter is shown in the following figure.

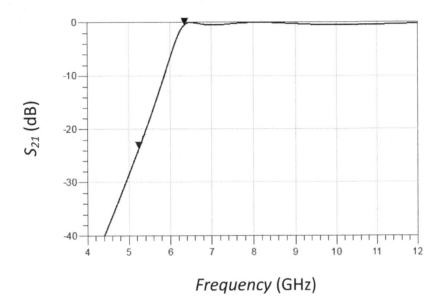

Problem 7. Design low-pass lumped-element filters with the following specifications:

	Filter 1 : 3 dB equal-ripple	Filter 2 : 0.5 dB equal-ripple
Cutoff frequency	$f_c = 2.24$ GHz	$f_c = 5.4$ GHz
Attenuation	$\alpha > 30$ dB at 3.36 GHz	$\alpha > 20$ dB at 6.5 GHz
Characteristic impedance	$Z_0 = 50\ \Omega$	$Z_0 = 75\ \Omega$

SOLUTION

Filter 1:

$$\left|\frac{\omega}{\omega_c}\right| - 1 = \frac{3.36}{2.24} - 1 = 0.5$$

- From the "Attenuation versus normalized frequency for equal-ripple filter prototypes (3 dB ripple level)" figure, find $N = 5$ for $\alpha > 30$ dB.

- From the "Element values for equal-ripple low-pass filter prototypes (3 dB ripple)" table, find:

$g_1 = 3.4817$

$g_2 = 0.7618$

$g_3 = 4.5381$

$g_4 = 0.7618$

$g_5 = 3.4817$

The low-pass filter:

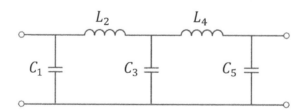

$$C_1 = \frac{g_1}{R_0 \omega_c} = 4.948 \text{ pF}$$

$$L_2 = \frac{R_0 g_2}{\omega_c} = 2.706 \text{ nH}$$

$$C_3 = \frac{g_3}{R_0 \omega_c} = 6.449 \text{ pF}$$

$$L_4 = \frac{R_0 g_4}{\omega_c} = 2.706 \text{ nH}$$

$$C_5 = \frac{g_5}{R_0 \omega_c} = 4.948 \text{ pF}$$

Note: The frequency response of this filter is shown in the following figure.

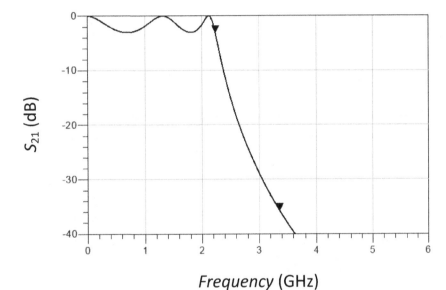

Filter 2:

$$\left|\frac{\omega}{\omega_c}\right| - 1 = 0.204$$

- From the "Attenuation versus normalized frequency for equal-ripple filter prototypes (0.5 dB ripple level)" figure, find $N = 7$ for $\alpha > 20$ dB.

- From the "Element values for equal-ripple low-pass filter prototypes (0.5 dB ripple)" table, find:

$g_1 = 1.7372$

$g_2 = 1.2583$

$g_3 = 2.6381$

$g_4 = 1.3444$

$g_5 = 2.6381$

$g_6 = 1.2583$

$g_7 = 1.7372$

The low-pass filter:

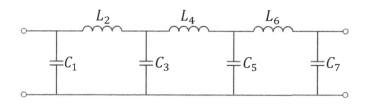

$C_1 = \dfrac{g_1}{R_0\omega_c} = 0.683$ pF

$L_2 = \dfrac{R_0 g_2}{\omega_c} = 2.781$ nH

$C_3 = \dfrac{g_3}{R_0\omega_c} = 1.037$ pF

$L_4 = \dfrac{R_0 g_4}{\omega_c} = 2.972$ nH

$C_5 = \dfrac{g_5}{R_0\omega_c} = 1.037$ pF

$$L_6 = \frac{R_0 g_6}{\omega_c} = 2.781 \text{ nH}$$

$$C_7 = \frac{g_7}{R_0 \omega_c} = 0.683 \text{ pF}$$

Note: The frequency response of this filter is shown in the following figure.

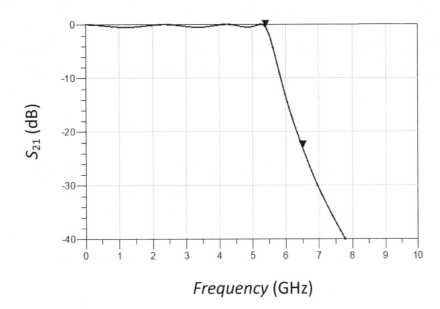

Problem 8. Design bandpass lumped-element filters with the following specifications:

	Filter 1 : 3 dB equal-ripple	Filter 2 : 0.5 dB equal-ripple
Order	$N = 3$	$N = 5$
Center frequency	$f_0 = 1.5$ GHz	$f_0 = 2.3$ GHz
Bandwidth	15 %	10 %
Impedance	$Z_0 = 50 \ \Omega$	$Z_0 = 75 \ \Omega$

SOLUTION

Low-pass to bandpass transformation:

Filter 1:

- From the "Element values for equal-ripple low-pass filter prototypes (3 dB ripple)" table, find:

$$g_1 = 3.3487$$

$$g_2 = 0.7117$$

$$g_3 = 3.3487$$

The bandpass filter:

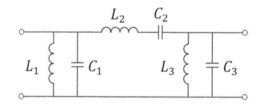

$$L_1 = \frac{\Delta Z_0}{\omega_0 g_1} = 0.238 \text{ nH}$$

$$C_1 = \frac{g_1}{\omega_0 \Delta Z_0} = 47.374 \text{ pF}$$

$$L_2 = \frac{g_2 Z_0}{\omega_0 \Delta} = 25.171 \text{ nH}$$

$$C_2 = \frac{\Delta}{\omega_0 g_2 Z_0} = 0.447 \text{ pF}$$

$$L_3 = \frac{\Delta Z_0}{\omega_0 g_3} = 0.238 \text{ nH}$$

$$C_3 = \frac{g_3}{\omega_0 \Delta Z_0} = 47.374 \text{ pF}$$

Note: The frequency response of this filter is shown in the following figure.

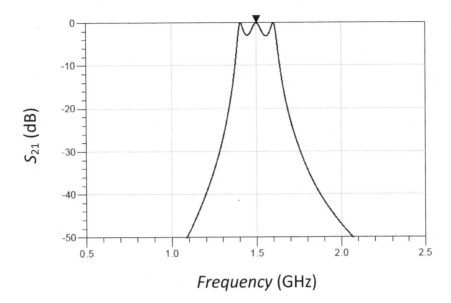

Filter 2:

- From the "Element values for equal-ripple low-pass filter prototypes (0.5 dB ripple)" table, find:

$g_1 = 1.7058$

$g_2 = 1.2296$

$g_3 = 2.5408$

$g_4 = 1.2296$

$g_5 = 1.7058$

The bandpass filter:

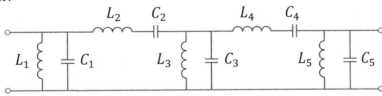

$L_1 = \dfrac{\Delta Z_0}{\omega_0 g_1} = 0.304 \text{ nH}$

$C_1 = \dfrac{g_1}{\omega_0 \Delta Z_0} = 15.738 \text{ pF}$

$$L_2 = \frac{g_2 Z_0}{\omega_0 \Delta} = 63.814 \text{ nH}$$

$$C_2 = \frac{\Delta}{\omega_0 g_2 Z_0} = 0.075 \text{ pF}$$

$$L_3 = \frac{\Delta Z_0}{\omega_0 g_3} = 0.204 \text{ nH}$$

$$C_3 = \frac{g_3}{\omega_0 \Delta Z_0} = 23.442 \text{ pF}$$

$$L_4 = \frac{g_4 Z_0}{\omega_0 \Delta} = 63.814 \text{ nH}$$

$$C_4 = \frac{\Delta}{\omega_0 g_4 Z_0} = 0.075 \text{ pF}$$

$$L_5 = \frac{\Delta Z_0}{\omega_0 g_5} = 0.304 \text{ nH}$$

$$C_5 = \frac{g_5}{\omega_0 \Delta Z_0} = 15.738 \text{ pF}$$

Note: The frequency response of this filter is shown in the following figure.

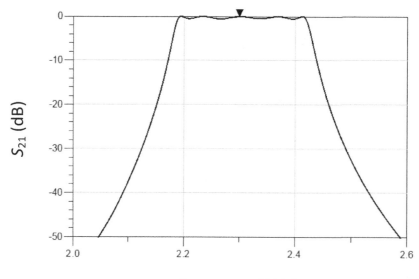

Frequency (GHz)

Problem 9. Design coupled line bandpass filters with the following specifications and find their attenuation at 1.5 GHz:

	Filter 1 : 3 dB equal-ripple	Filter 2 : 0.5 dB equal-ripple
Order	$N=3$	$N=3$
Center frequency	$f_0 = 1.74$ GHz	$f_0 = 2.1$ GHz
Bandwidth	10 %	15 %
Impedance	$Z_0 = 50\ \Omega$	$Z_0 = 75\ \Omega$

SOLUTION

Low-pass to bandpass transformation:

Filter 1:

$$\omega \leftarrow \frac{1}{\Delta}\left(\frac{\omega}{\omega_0} - \frac{\omega_0}{\omega}\right) = \frac{1}{0.1}\left(\frac{1.5}{1.74} - \frac{1.74}{1.5}\right) = -2.979$$

$$\left|\frac{\omega}{\omega_c}\right| - 1 = |-2.979| - 1 = 1.979$$

- From the "Attenuation versus normalized frequency for equal-ripple filter prototypes (3 dB ripple level)" figure, the attenuation is around 38 dB for $N= 3$.

- From the "Element values for equal-ripple low-pass filter prototypes (3 dB ripple)" table, find:

$$g_1 = 3.3487$$

$$g_2 = 0.7117$$

$$g_3 = 3.3487$$

The admittance inverter constants J_n are given by:

$$Z_0 J_1 = \sqrt{\frac{\pi\Delta}{2g_1}}$$

$$Z_0 J_n = \frac{\pi\Delta}{2\sqrt{g_{n-1} \times g_n}} \qquad \text{for} \qquad n = 2, 3, ..., N$$

$$Z_0 J_{N+1} = \sqrt{\frac{\pi\Delta}{2g_N \times g_{N+1}}}$$

The even- and odd-mode characteristic impedances for each section are given by:

$$Z_{0e} = Z_0 \left[1 + J Z_0 + (J Z_0)^2 \right]$$

$$Z_{0o} = Z_0 \left[1 - J Z_0 + (J Z_0)^2 \right]$$

All the results are summarized in the following table:

n	g_n	$Z_0 J_n$	$Z_{0e}(\Omega)$	$Z_{0o}(\Omega)$
1	3.3487	0.2166	63.18	41.52
2	0.7117	0.1017	55.60	45.43
3	3.3487	0.1017	55.60	45.43
4	1.0000	0.2166	63.18	41.52

The bandpass filter:

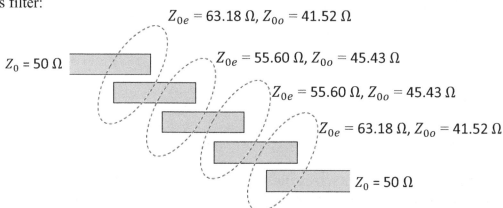

The length of all the lines is $\lambda/4$ at $f = 1.74$ GHz.

Note: The frequency response of this filter is shown in the following figure.

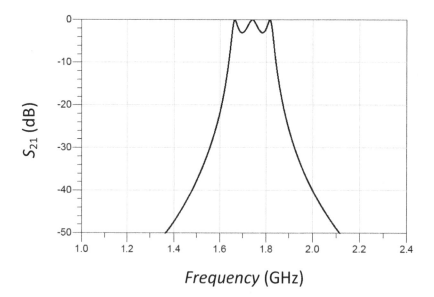

Filter 2:

$$\omega \leftarrow \frac{1}{\Delta}(\frac{\omega}{\omega_0} - \frac{\omega_0}{\omega}) = -4.571$$

$$\left|\frac{\omega}{\omega_c}\right| - 1 = 3.571$$

- From the "Attenuation versus normalized frequency for equal-ripple filter prototypes (0.5 dB ripple level)" figure, the attenuation is around 38 dB for $N = 3$.

- From the "Element values for equal-ripple low-pass filter prototypes (0.5 dB ripple)" table, find:

$g_1 = 1.5963$

$g_2 = 1.0967$

$g_3 = 1.5963$

All the results are summarized in the following table:

n	g_n	$Z_0 J_n$	$Z_{0e}(\Omega)$	$Z_{0o}(\Omega)$
1	1.5963	0.3842	114.89	57.26
2	1.0967	0.1781	90.74	64.02
3	1.5963	0.1781	90.74	64.02
4	1.0000	0.3842	114.89	57.26

The bandpass filter:

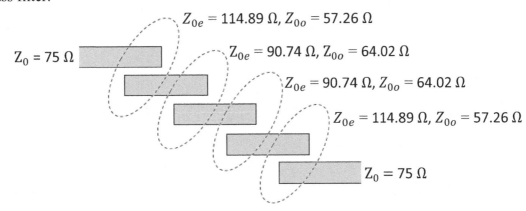

The length of all the lines is $\lambda/4$ at $f = 2.1$ GHz.

Note: The frequency response of this filter is shown in the following figure.

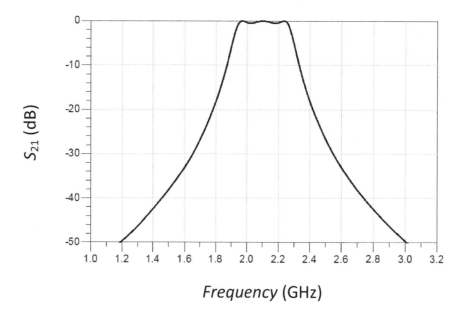

PART THREE
IMPEDANCE MATCHING AND TUNING

Problem 1. Design lossless L-section matching circuits to match the loads to the generators at the given frequencies:

Load	Generator's impedance	Frequency
$Z_L = 240 - j120\ \Omega$	$150\ \Omega$	5 GHz
$Z_L = 10 + j40\ \Omega$	$100\ \Omega$	2 GHz
$Z_L = 15 - j45\ \Omega$	$75\ \Omega$	3 GHz
$Z_L = 40 - j15\ \Omega$	$50\ \Omega$	1.5 GHz

SOLUTION

1. For $Z_L = 240 - j120\ \Omega$, $Z_0 = 150\ \Omega$ and $f = 5$ GHz:

Step 1: Normalize $Z_L \rightarrow z_L = Z_L / Z_0$

$$z_L = \frac{240 - j120}{150} = 1.6 - j0.8$$

Step 2: Locate and plot z_L on the Smith chart.

Since z_L is inside the $1 + jx$ circle, the following matching circuit is used:

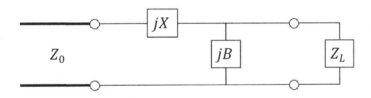

Step 3: Since the first element from the load is a shunt susceptance, convert to admittance and find $y_L = 0.5 + j0.25$.

Step 4: Construct the $1 + jx$ circle on the admittance Smith chart.

Step 5: Add a shunt susceptance to move from y_L to the rotated $1 + jx$ circle. Find that $jb = j0.25$ will move y_L along a constant-conductance circle to $y = 0.5 + j0.5$.

Note: This first solution is the shortest distance from y_L to the rotated $1 + jx$ circle.

Step 6: Convert back y to impedance. Find that $z = 1 - j1$.

Step 7: Add a series reactance to move z to the center of the chart. Find that $jx = j1$.

Note: Notice that b and x have positive values and this matching circuit consists (starting from the load) of a shunt susceptance followed by a series reactance. Therefore, this matching circuit consists of a shunt capacitor and a series inductor.

The capacitor value is:

$$C = \frac{b}{2\pi f Z_0} = 0.053 \text{ pF}$$

The inductor value is:

$$L = \frac{x Z_0}{2\pi f} = 4.775 \text{ nH}$$

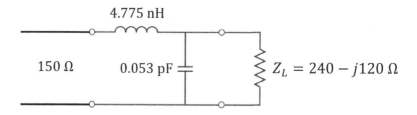

The second solution is (the longest distance from y_L to the rotated $1 + jx$ circle):

$$b = -0.75$$

$$x = -1$$

Note: Notice that b and x have negative values and this matching circuit consists (starting from the load) of a shunt susceptance followed by a series reactance. Therefore, this matching circuit consists of a shunt inductor and a series capacitor.

The inductor value is:

$$L = \frac{-Z_0}{2\pi f b} = 6.366 \text{ nH}$$

The capacitor value is:

$$C = \frac{-1}{2\pi f x Z_0} = 0.212 \text{ pF}$$

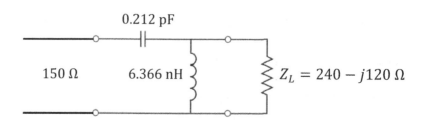

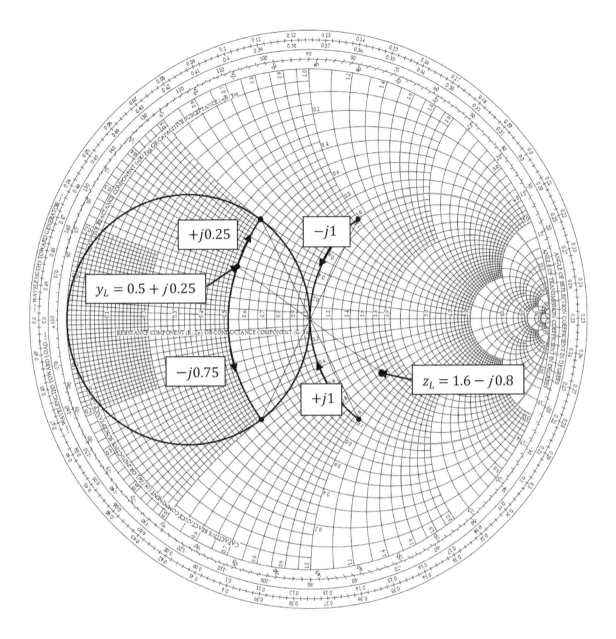

2. For $Z_L = 10 + j40 \ \Omega$, $Z_0 = 100 \ \Omega$ and $f = 2$ GHz:

$z_L = 0.1 + j0.4$

Since z_L is outside the $1 + jx$ circle, the following matching circuit is used:

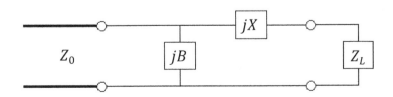

Note: The first element from the load is a series reactance. Thus, add a series reactance to move from z_L to the rotated $1+jx$ circle before converting to admittance.

The first solution:

$$x = -0.1$$

$$b = +3$$

Notice that x is negative, b is positive and this matching circuit consists (starting from the load) of a series reactance followed by a shunt susceptance. Therefore, this matching circuit consists of a series capacitor and a shunt capacitor.

The series capacitor value is:

$$C = \frac{-1}{2\pi f x Z_0} = 7.958 \text{ pF}$$

The shunt capacitor value is:

$$C = \frac{b}{2\pi f Z_0} = 2.387 \text{ pF}$$

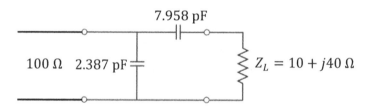

100 Ω   2.387 pF   7.958 pF $Z_L = 10 + j40\ \Omega$

The second solution:

$$x = -0.7$$

$$b = -3$$

Notice that x and b have negative values and this matching circuit consists (starting from the load) of a series reactance followed by a shunt susceptance. Therefore, this matching circuit consists of a series capacitor and a shunt inductor.

The capacitor value is:

$$C = \frac{-1}{2\pi f x Z_0} = 1.137 \text{ pF}$$

The inductor value is:

$$L = \frac{-Z_0}{2\pi f b} = 2.653 \text{ nH}$$

74

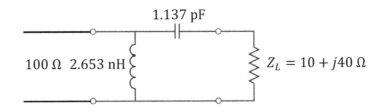

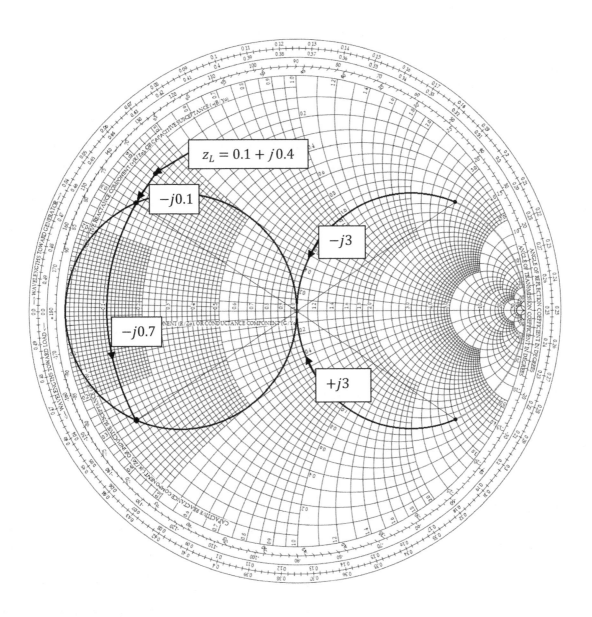

3. For $Z_L = 15 - j45\ \Omega$, $Z_0 = 75\ \Omega$ and $f = 3$ GHz:

$z_L = 0.2 - j0.6$

Since z_L is outside the $1+jx$ circle, the following matching circuit is used:

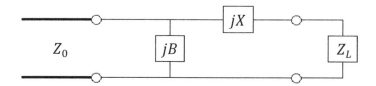

The first solution:

$x = +0.2$

$b = -2$

Notice that x is positive, b is negative and this matching circuit consists (starting from the load) of a series reactance followed by a shunt susceptance. Therefore, this matching circuit consists of a series inductor and a shunt inductor.

The series inductor value is:

$$L = \frac{xZ_0}{2\pi f} = 0.796\ \text{nH}$$

The shunt inductor value is:

$$L = \frac{-Z_0}{2\pi f b} = 1.989\ \text{nH}$$

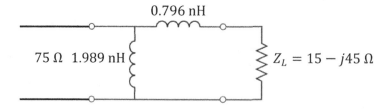

The second solution:

$x = +1$

$b = +2$

Notice that x and b have positive values and this matching circuit consists (starting from the load) of a series reactance followed by a shunt susceptance. Therefore, this matching circuit consists of a series inductor and a shunt capacitor.

The inductor value is:

$$L = \frac{xZ_0}{2\pi f} = 3.979 \text{ nH}$$

The capacitor value is:

$$C = \frac{b}{2\pi f Z_0} = 1.415 \text{ pF}$$

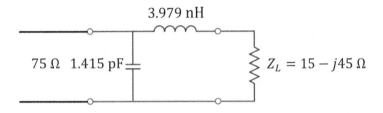

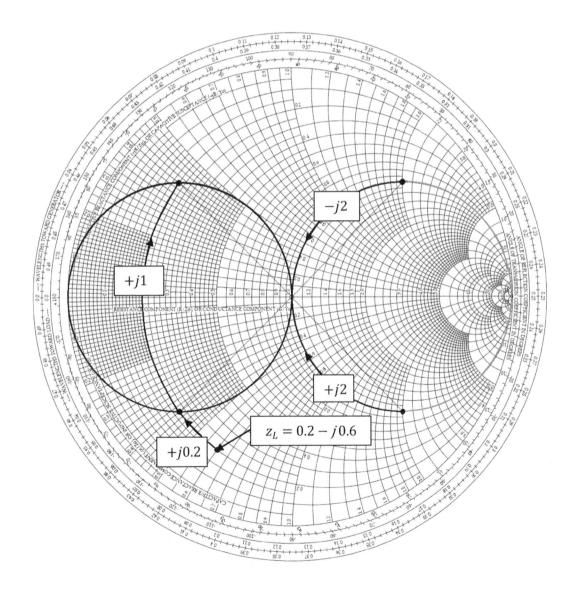

4. For $Z_L = 40 - j15 \ \Omega$, $Z_0 = 50 \ \Omega$ and $f = 1.5$ GHz:

$z_L = 0.8 - j0.3$

Since z_L is outside the $1 + jx$ circle, the following matching circuit is used:

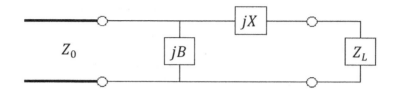

The first solution:

$x = -0.1$

$b = -0.5$

Notice that x and b have negative values and this matching circuit consists (starting from the load) of a series reactance followed by a shunt susceptance. Therefore, this matching circuit consists of a series capacitor and a shunt inductor.

The capacitor value is:

$$C = \frac{-1}{2\pi f x Z_0} = 21.221 \text{ pF}$$

The inductor value is:

$$L = \frac{-Z_0}{2\pi f b} = 10.610 \text{ nH}$$

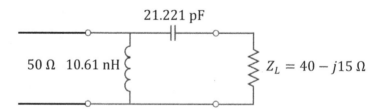

The second solution:

$x = +0.7$

$b = +0.5$

Notice that x and b have positive values and this matching circuit consists (starting from the load) of a series reactance followed by a shunt susceptance. Therefore, this matching circuit consists of a series inductor and a shunt capacitor.

The inductor value is:

$$L = \frac{xZ_0}{2\pi f} = 3.714 \text{ nH}$$

The capacitor value is:

$$C = \frac{b}{2\pi f Z_0} = 1.061 \text{ pF}$$

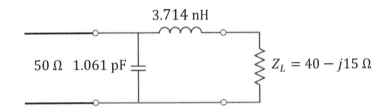

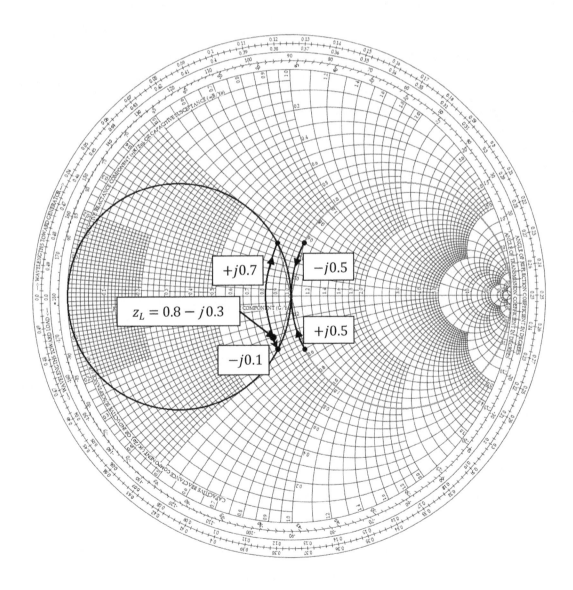

Problem 2. Using a single shunt open-circuit stub tuner, find two designs to match each load to its corresponding line:

Load	Line's impedance
$Z_L = 120 - j90 \ \Omega$	$75 \ \Omega$
$Z_L = 100 + j50 \ \Omega$	$50 \ \Omega$

SOLUTION

1. For $Z_L = 120 - j90 \ \Omega$, $Z_0 = 75 \ \Omega$:

Step 1: Normalize $Z_L \rightarrow z_L = Z_L / Z_0$

$$z_L = \frac{120 - j90}{75} = 1.6 - j1.2$$

Step 2: Locate and plot z_L on the Smith chart.

Step 3: Since a shunt-stub is to be used, construct the SWR circle and convert to the load admittance. Find $y_L = 0.4 + j0.3$.

Note: The SWR circle intersects the $1 + jb$ circle at two points, denoted as y_1 and y_2. The distance d from the load to the stub is given by either of these two intersections.

Step 4: On the *WAVELENGHTS TOWARD GENERATOR* scale, read: 0.054λ for y_L, 0.164λ for y_1 and 0.337λ for y_2.

Then, the first distance is: $\quad d_1 = 0.164\lambda - 0.054\lambda = 0.110\lambda$

The second distance is: $\quad d_2 = 0.337\lambda - 0.054\lambda = 0.283\lambda$

Step 5: At the two intersection points, find $y_1 = 1 + j1.06$ and $y_2 = 1 - j1.06$.

Step 6: The first tuning solution requires a stub with a susceptance of $-j1.06$. Find the length of an open-circuited stub that gives this susceptance by starting at $y = 0$ (the open circuit) and moving along the outer edge of the chart ($g = 0$) toward the generator to the $-j1.06$ point.

The first stub length is then: $\ell_1 = 0.371\lambda$

Similarly, the second stub length is: $\ell_2 = 0.129\lambda$

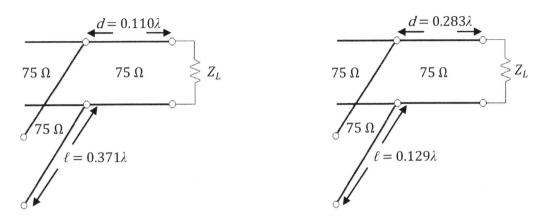

-First solution-   -Second solution-

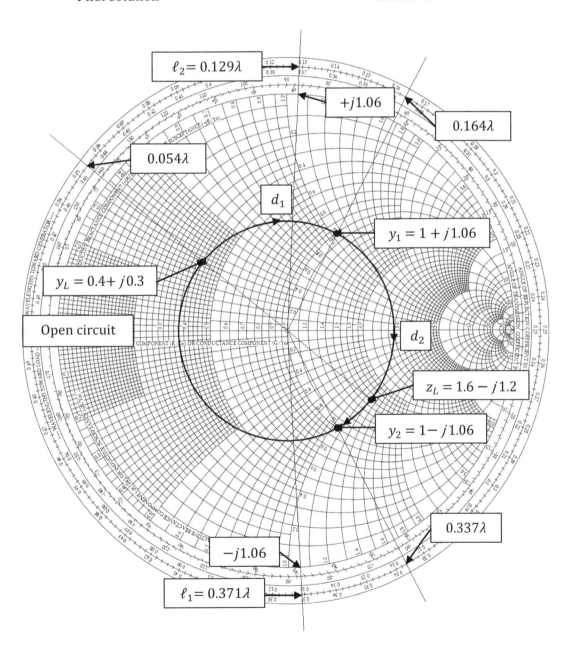

Note: These results can be found analytically as follows:

$$t = \frac{X_L \pm \sqrt{R_L[(Z_0 - R_L)^2 + X_L^2]/Z_0}}{R_L - Z_0}$$

$$t = \frac{-90 \pm \sqrt{120[(75-120)^2 + (-90)^2]/75}}{120 - 75} \rightarrow t_1 = 0.828 \text{ and } t_2 = -4.828$$

The possible stub positions are:

$$d_1 = \frac{\lambda}{2\pi} \tan^{-1} t_1 = 0.110\lambda$$

$$d_2 = \frac{\lambda}{2\pi} (\pi + \tan^{-1} t_2) = 0.283\lambda$$

The required stub susceptances are:

$$B = \frac{R_L^2 t - (Z_0 - X_L t)(X_L + Z_0 t)}{Z_0[R_L^2 + (X_L + Z_0 t)^2]} \rightarrow B_1 = 0.014 \text{ and } B_2 = -0.014$$

The open-circuited stub lengths are:

$$\ell_1 = \frac{-\lambda}{2\pi} \tan^{-1}(B_1 Z_0) = 0.371\lambda \ (\lambda/2 \text{ is added to get } \ell_1 > 0)$$

$$\ell_2 = \frac{-\lambda}{2\pi} \tan^{-1}(B_2 Z_0) = 0.129\lambda$$

2. For $Z_L = 100 + j50 \ \Omega$, $Z_0 = 50 \ \Omega$:

$z_L = 2 + j1 \rightarrow y_L = 0.4 - j0.2$

At the two intersection points: $y_1 = 1 + j1$ and $y_2 = 1 - j1$

The first distance: $d_1 = 0.162\lambda + (0.5 - 0.463) \lambda = 0.199\lambda$

The second distance: $d_2 = 0.338\lambda + (0.5 - 0.463) \lambda = 0.375\lambda$

The first stub length: $\ell_1 = 0.375\lambda$

The second stub length: $\ell_2 = 0.125\lambda$

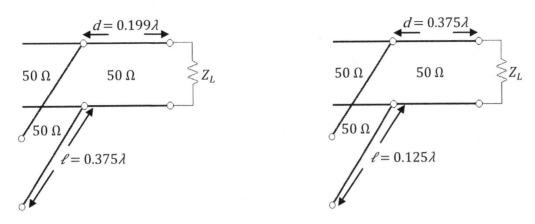

-First solution-   -Second solution-

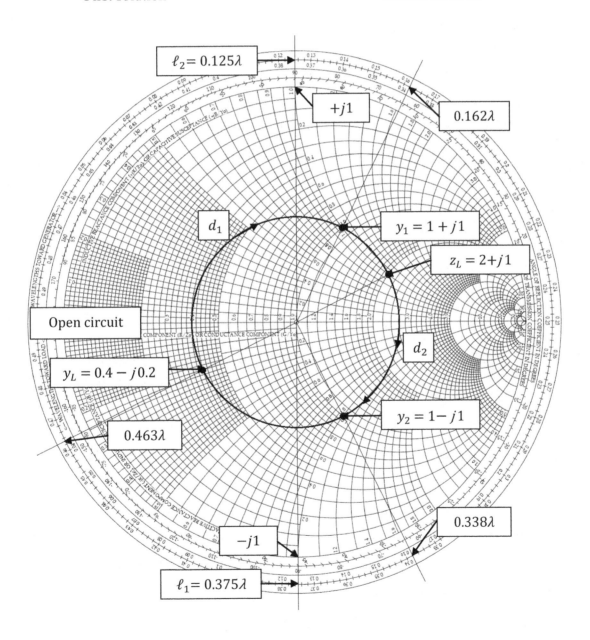

Note: Analytically:

$$t = \frac{X_L \pm \sqrt{R_L[(Z_0 - R_L)^2 + X_L^2]/Z_0}}{R_L - Z_0}$$

$$t = \frac{50 \pm \sqrt{100[(50-100)^2 + (50)^2]/50}}{100 - 50} \rightarrow t_1 = 3 \text{ and } t_2 = -1$$

The possible stub positions are:

$$d_1 = \frac{\lambda}{2\pi} \tan^{-1} t_1 = 0.199\lambda$$

$$d_2 = \frac{\lambda}{2\pi} (\pi + \tan^{-1} t_2) = 0.375\lambda$$

The required stub susceptances are:

$$B = \frac{R_L^2 t - (Z_0 - X_L t)(X_L + Z_0 t)}{Z_0[R_L^2 + (X_L + Z_0 t)^2]} \rightarrow B_1 = 0.02 \text{ and } B_2 = -0.02$$

The open-circuited stub lengths are:

$$\ell_1 = \frac{-\lambda}{2\pi} \tan^{-1}(B_1 Z_0) = 0.375\lambda \, (\lambda/2 \text{ is added to get } \ell_1 > 0)$$

$$\ell_2 = \frac{-\lambda}{2\pi} \tan^{-1}(B_2 Z_0) = 0.125\lambda$$

Problem 3. Repeat the previous problem using short-circuited stubs.

SOLUTION

1. For $Z_L = 120 - j90 \, \Omega$, $Z_0 = 75 \, \Omega$:

The first distance: $d_1 = 0.110\lambda$

The second distance: $d_2 = 0.283\lambda$

The length of a short-circuited stub that gives a susceptance of $-j1.06$ is determined by starting at $y = \infty$ and moving along the outer edge of the chart toward the generator to the $-j1.06$ point.

The first stub length is then: $\ell_1 = 0.371\lambda - 0.25\lambda = 0.121\lambda$

Similarly, the second stub length is: $\ell_2 = 0.25\lambda + 0.129\lambda = 0.379\lambda$

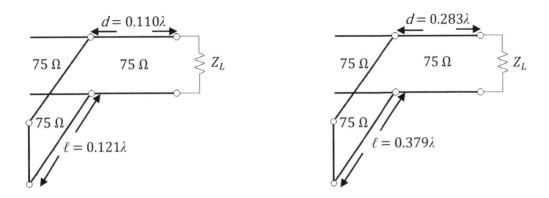

-First solution-   -Second solution-

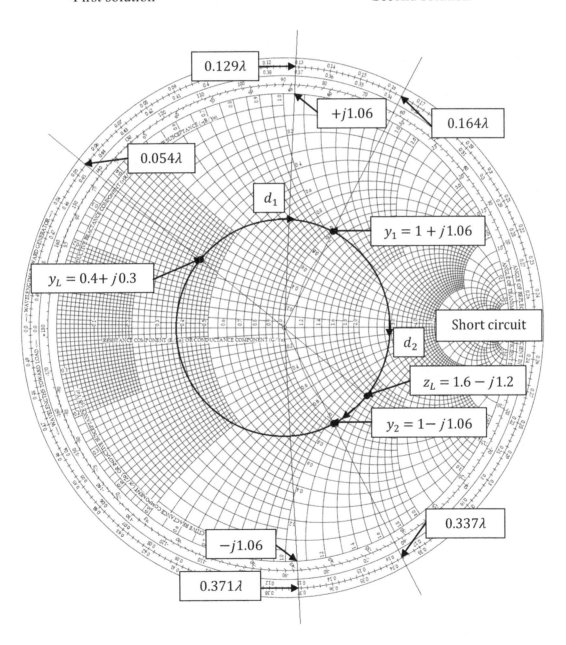

Note: Lengths of the short-circuited stubs can be found analytically as follows:

$$\ell_1 = \frac{\lambda}{2\pi} \, tan^{-1} \frac{1}{B_1 Z_0} = 0.121\lambda$$

$$\ell_2 = \frac{\lambda}{2\pi} \, tan^{-1} \frac{1}{B_2 Z_0} = 0.379\lambda \ (\lambda/2 \text{ is added to get } \ell_2 > 0)$$

2. For $Z_L = 100 + j50 \ \Omega$, $Z_0 = 50 \ \Omega$:

The first distance: $d_1 = 0.199\lambda$

The second distance: $d_2 = 0.375\lambda$

The first stub length: $\ell_1 = 0.375\lambda - 0.25\lambda = 0.125\lambda$

The second stub length: $\ell_2 = 0.25\lambda + 0.125\lambda = 0.375\lambda$

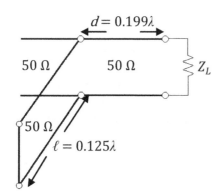

-First solution-

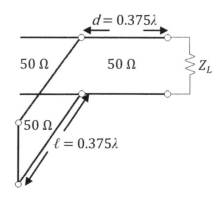

-Second solution-

Note: Analytically:

$$\ell_1 = \frac{\lambda}{2\pi} \, tan^{-1} \frac{1}{B_1 Z_0} = 0.125\lambda$$

$$\ell_2 = \frac{\lambda}{2\pi} \, tan^{-1} \frac{1}{B_2 Z_0} = 0.375\lambda \ (\lambda/2 \text{ is added to get } \ell_2 > 0)$$

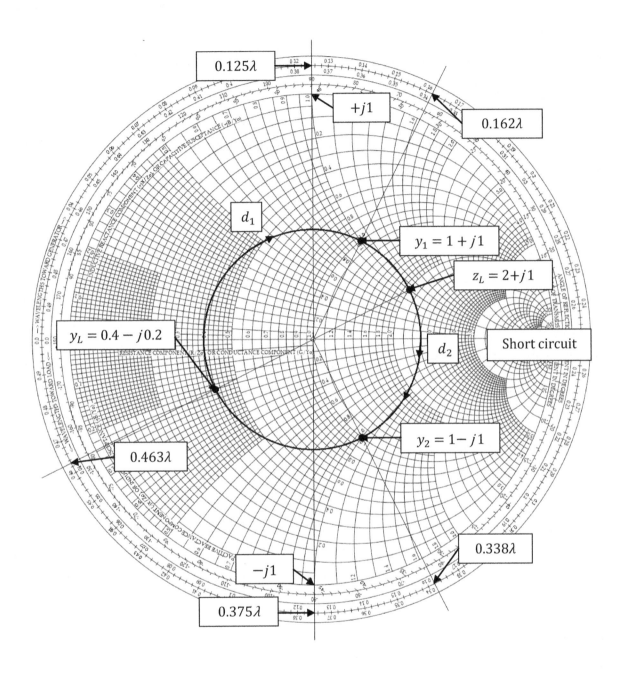

Problem 4. Using a single series open-circuit stub tuner, find two designs to match each load to its corresponding line:

Load	Line's impedance
$Z_L = 135 - j75 \, \Omega$	$75 \, \Omega$
$Z_L = 80 + j40 \, \Omega$	$50 \, \Omega$

SOLUTION

1. For $Z_L = 135 - j75 \, \Omega$, $Z_0 = 75 \, \Omega$:

$$z_L = \frac{135 - j75}{75} = 1.8 - j1$$

Note: Since a series stub is to be used, construct the SWR circle without converting to the load admittance.

The SWR circle intersects the $1 + jx$ circle at two points, find $z_1 = 1 - j0.95$ and $z_2 = 1 + j0.95$.

On the *WAVELENGHTS TOWARD GENERATOR* scale, read: 0.294λ for z_L, 0.34λ for z_1 and 0.16λ for z_2.

Then, the first distance is: $d_1 = 0.34\lambda - 0.294\lambda = 0.046\lambda$

The second distance is: $d_2 = 0.16\lambda + (0.5 - 0.294)\,\lambda = 0.366\lambda$

The first tuning solution requires a stub with a reactance of $+j0.95$. Find the length of an open-circuited stub that gives this reactance by starting at $z = \infty$ (the open circuit) and moving along the outer edge of the chart ($r = 0$) toward the generator to the $+j0.95$ point.

The first stub length is then: $\ell_1 = 0.25\lambda + 0.121\lambda = 0.371\,\lambda$

Similarly, the second stub length is: $\ell_2 = 0.379\lambda - 0.25\lambda = 0.129\,\lambda$

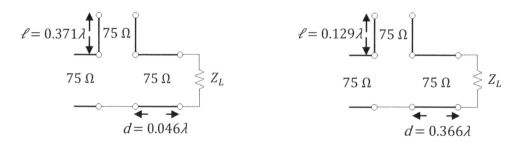

-First solution-   -Second solution-

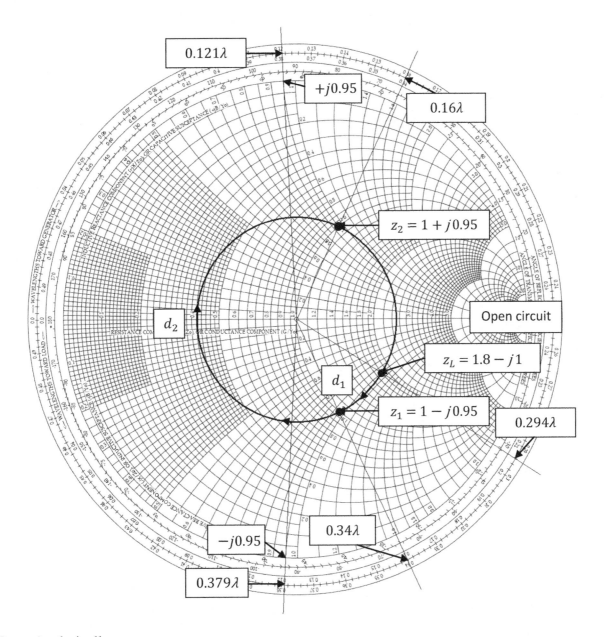

Note: Analytically:

$$t = \frac{B_L \pm \sqrt{G_L[(Y_0 - G_L)^2 + B_L{}^2]/Y_0}}{G_L - Y_0} \text{ with } Y_L = 1/Z_L = G_L + jB_L$$

$\rightarrow t_1 = 0.2943$ and $t_2 = -1.1140$

The possible stub positions are:

$$d_1 = \frac{\lambda}{2\pi} \tan^{-1} t_1 = 0.046\lambda$$

$$d_2 = \frac{\lambda}{2\pi} (\pi + \tan^{-1} t_2) = 0.366\lambda$$

The reactance X :

$$X = \frac{G_L^2 t - (Y_0 - tB_L)(B_L + tY_0)}{Y_0[G_L^2 + (B_L + Y_0 t)^2]} \rightarrow X_1 = -71.59 \text{ and } X_2 = 71.59$$

The open-circuited stub lengths are:

$$\ell_1 = \frac{\lambda}{2\pi} \tan^{-1}\left(\frac{Z_0}{X_1}\right) = 0.371\lambda \ (\lambda/2 \text{ is added to get } \ell_1 > 0)$$

$$\ell_2 = \frac{\lambda}{2\pi} \tan^{-1}\left(\frac{Z_0}{X_2}\right) = 0.129\lambda$$

2. For $Z_L = 80 + j40 \ \Omega$, $Z_0 = 50 \ \Omega$:

$z_L = 1.6 + j0.8$

At the two intersection points: $z_1 = 1 - j0.79$ and $z_2 = 1 + j0.79$

The first distance is: $d_1 = 0.345\lambda - 0.2\lambda = 0.145\lambda$

The second distance is: $d_2 = 0.155\lambda + (0.5 - 0.2)\lambda = 0.455\lambda$

The first stub length is: $\ell_1 = 0.25\lambda + 0.106\lambda = 0.356\lambda$

The second stub length is: $\ell_2 = 0.394\lambda - 0.25\lambda = 0.144\lambda$

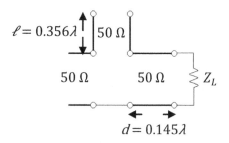

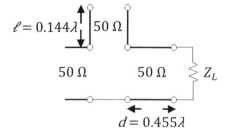

-First solution-   -Second solution-

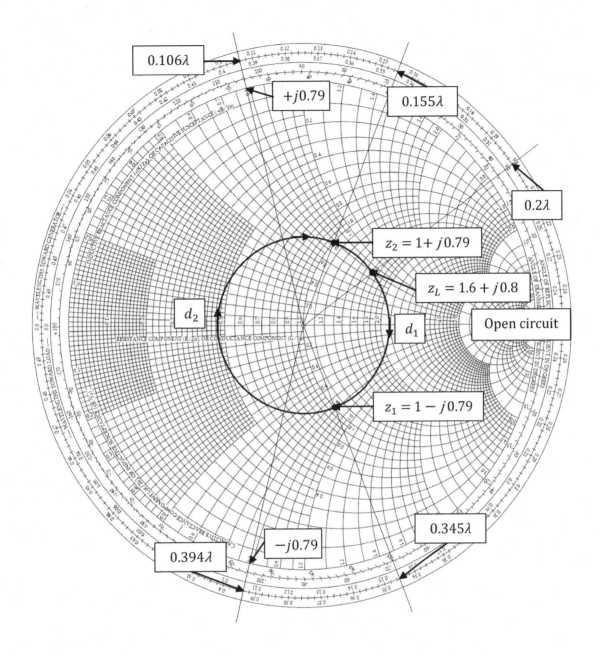

Note: Analytically:

$$t = \frac{B_L \pm \sqrt{G_L[(Y_0 - G_L)^2 + B_L{}^2]/Y_0}}{G_L - Y_0} \rightarrow t_1 = 1.2906 \text{ and } t_2 = -0.2906.$$

The possible stub positions are:

$$d_1 = \frac{\lambda}{2\pi} \, tan^{-1} t_1 = 0.145\lambda$$

$$d_2 = \frac{\lambda}{2\pi} \, (\pi + tan^{-1} t_2) = 0.455\lambda$$

The reactance X :

$$X = \frac{G_L{}^2 t - (Y_0 - tB_L)(B_L + tY_0)}{Y_0[G_L{}^2 + (B_L + Y_0 t)^2]} \rightarrow X_1 = -39.53 \text{ and } X_2 = 39.53$$

The open-circuited stub lengths are:

$$\ell_1 = \frac{\lambda}{2\pi} \tan^{-1}(\frac{Z_0}{X_1}) = 0.356\lambda \ (\lambda/2 \text{ is added to get } \ell_1 > 0)$$

$$\ell_2 = \frac{\lambda}{2\pi} \tan^{-1}(\frac{Z_0}{X_2}) = 0.144\lambda$$

Problem 5. Repeat the previous problem using short-circuited stubs.

SOLUTION

1. For $Z_L = 135 - j75 \ \Omega$, $Z_0 = 75 \ \Omega$:

The first distance: $d_1 = 0.046\lambda$

The second distance: $d_2 = 0.366\lambda$

The length of a short-circuited stub that gives a reactance of $+j0.95$ is determined by starting at $z = 0$ (the short circuit) and moving along the outer edge of the chart toward the generator to the $+j$ 0.95 point.

The first stub length is then: $\ell_1 = 0.121\lambda$

Similarly, the second stub length is: $\ell_2 = 0.379\lambda$

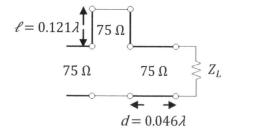

-First solution-   -Second solution-

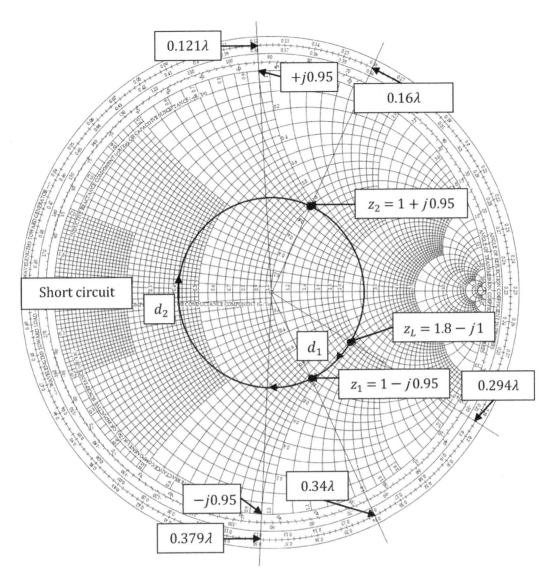

Note: Lengths of the short-circuited stubs can be found analytically as follows:

$$\ell_1 = \frac{-\lambda}{2\pi}\,tan^{-1}(\frac{X_1}{Z_0}) = 0.121\lambda$$

$$\ell_2 = \frac{-\lambda}{2\pi}\,tan^{-1}(\frac{X_2}{Z_0}) = 0.379\lambda\ (\lambda/2 \text{ is added to get } \ell_2 > 0)$$

2. For $Z_L = 80 + j40\ \Omega$, $Z_0 = 50\ \Omega$:

The first distance: $d_1 = 0.145\lambda$

The second distance: $d_2 = 0.455\lambda$

The first stub length: $\ell_1 = 0.106\lambda$

The second stub length: $\ell_2 = 0.394\lambda$

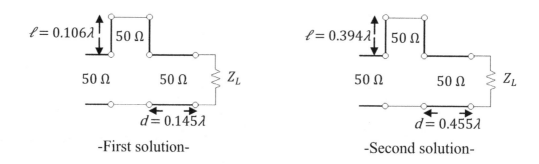

-First solution- -Second solution-

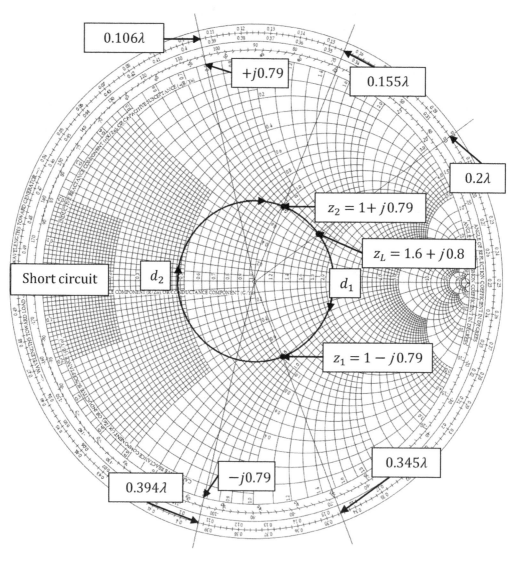

Note: Analytically:

$$\ell_1 = \frac{-\lambda}{2\pi} \, tan^{-1}\left(\frac{X_1}{Z_0}\right) = 0.106\lambda$$

$$\ell_2 = \frac{-\lambda}{2\pi} \, tan^{-1}\left(\frac{X_2}{Z_0}\right) = 0.394\lambda \; (\lambda/2 \text{ is added to get } \ell_2 > 0)$$

Problem 6. Design for each of the following cases a double-stub shunt tuner using open-circuited stubs with a $\lambda/8$ spacing to match the load to the line:

Load	Line's impedance
$Z_L = 93.75 - j93.75 \; \Omega$	$75 \; \Omega$
$Z_L = 25 - j75 \; \Omega$	$50 \; \Omega$

SOLUTION

1. For $Z_L = 93.75 - j93.75 \; \Omega$, $Z_0 = 75 \; \Omega$:

Step 1: Normalize $Z_L \rightarrow z_L = Z_L / Z_0$

$$z_L = \frac{93.75 - j\,93.75}{75} = 1.25 - j1.25$$

Step 2: Locate and plot z_L on the Smith chart.

Step 3: Convert to the load admittance. Find $y_L = 0.4 + j0.4$.

Step 4: Construct the rotated $1+jb$ conductance circle by moving every point on the $g = 1$ circle $\lambda/8$ toward the load.

Step 5: Move y_L to the rotated $1+jb$ circle. Find the susceptance of the first stub which can be one of two possible values: $b_1 = 1.4$ or $b'_1 = -0.2$.

Step 6: Move the two solutions $\lambda/8$ toward the generator. Find $y_2 = 1 - j3$ and $y'_2 = 1 + j1$. Then, the susceptance of the second stub (to move to the center of chart) can be one of two possible values: $b_2 = 3$ or $b'_2 = -1$.

Step 7: On the *WAVELENGHTS TOWARD GENERATOR* scale, read: 0.151λ for $+j1.4$, 0.469λ for $-j0.2$, 0.199λ for $+j3$ and 0.375λ for $-j1$.

Then, the lengths of the open-circuited stubs are:

$$\ell_1 = 0.151\lambda, \; \ell_2 = 0.199\lambda \qquad \text{or} \qquad \ell_1 = 0.469\lambda, \; \ell_2 = 0.375\lambda$$

-First solution-   -Second solution-

95

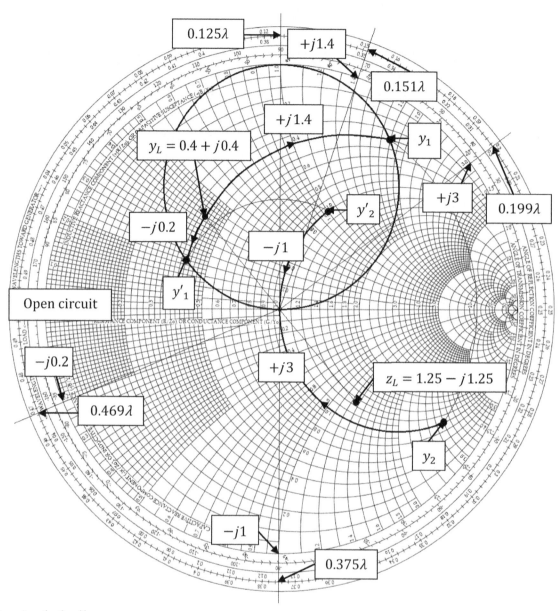

Note: Analytically:

$$t = tan(\beta d) = tan(\frac{2\pi}{\lambda} \times \frac{\lambda}{8}) = 1$$

$$b_1 = -b_L + \frac{1 \pm \sqrt{(1+t^2)g_L - g_L^2 t^2}}{t} \rightarrow b_1 = +1.4 \text{ or } b_1 = -0.2$$

$$b_2 = \frac{\pm \sqrt{(1+t^2)g_L - g_L^2 t^2} + g_L}{g_L t} \rightarrow b_2 = +3 \text{ or } b_2 = -1$$

$$\ell_1 = \frac{\lambda}{2\pi} tan^{-1}(b_1) = 0.151\lambda \text{ or } \ell_1 = 0.469\lambda$$

$$\ell_2 = \frac{\lambda}{2\pi} tan^{-1}(b_2) = 0.199\lambda \text{ or } \ell_2 = 0.375\lambda$$

2. For $Z_L = 25 - j75\ \Omega$, $Z_0 = 50\ \Omega$:

$z_L = 0.5 - j1.5 \rightarrow y_L = 0.2 + j0.6$

The susceptance of the first stub:

$b_1 = 1$ or $b'_1 = -0.2$

The susceptance of the second stub:

$b_2 = 4$ or $b'_2 = -2$

The lengths of the open-circuited stubs are:

$\ell_1 = 0.125\lambda,\ \ell_2 = 0.211\lambda$ or $\ell_1 = 0.469\lambda,\ \ell_2 = 0.324\lambda$

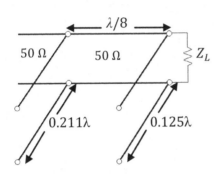
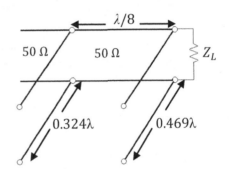

-First solution-   -Second solution-

Note: Analytically:

$t = tan(\beta d) = 1$

$b_1 = -b_L + \dfrac{1 \pm \sqrt{(1+t^2)g_L - g_L^2 t^2}}{t} \rightarrow b_1 = +1$ or $b_1 = -0.2$

$b_2 = \dfrac{\pm\sqrt{(1+t^2)g_L - g_L^2 t^2} + g_L}{g_L t} \rightarrow b_2 = +4$ or $b_2 = -2$

$\ell_1 = \dfrac{\lambda}{2\pi} tan^{-1}(b_1) = 0.125\lambda$ or $\ell_1 = 0.469\lambda$

$\ell_2 = \dfrac{\lambda}{2\pi} tan^{-1}(b_2) = 0.211\lambda$ or $\ell_2 = 0.324\lambda$

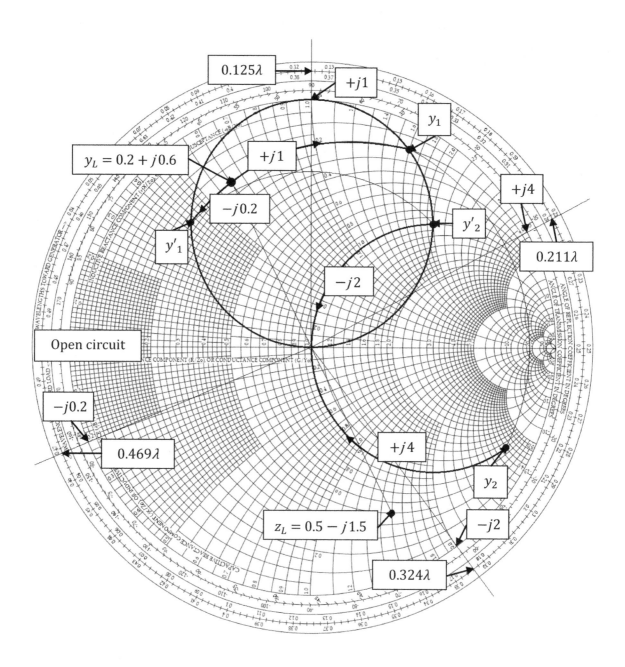

Problem 7. Repeat the previous problem using short-circuited stubs with a $3\lambda/8$ spacing.

SOLUTION

1. For $Z_L = 93.75 - j93.75 \ \Omega$, $Z_0 = 75 \ \Omega$:

$y_L = 0.4 + j0.4$

The susceptance of the first stub:

$b_1 = -2.2$ or $b'_1 = -0.6$

The susceptance of the second stub:

$b_2 = -3$ or $b'_2 = +1$

The lengths of the short-circuited stubs are:

$\ell_1 = 0.318\lambda - 0.25\lambda = 0.068\lambda$ or $\ell_1 = 0.414\lambda - 0.25\lambda = 0.164\lambda$

$\ell_2 = 0.301\lambda - 0.25\lambda = 0.051\lambda$ or $\ell_2 = 0.125\lambda + 0.25\lambda = 0.375\lambda$

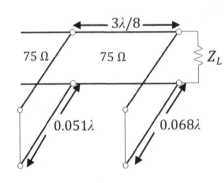 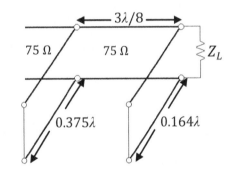

-First solution-   -Second solution-

Note: Analytically:

$$t = tan(\beta d) = tan(\frac{2\pi}{\lambda} \times \frac{3\lambda}{8}) = -1$$

$$b_1 = -b_L + \frac{1 \pm \sqrt{(1+t^2)g_L - g_L^2 t^2}}{t} \ \rightarrow b_1 = -2.2 \text{ or } b_1 = -0.6$$

$$b_2 = \frac{\pm\sqrt{(1+t^2)g_L - g_L^2 t^2} + g_L}{g_L t} \ \rightarrow b_2 = -3 \text{ or } b_2 = +1$$

$$\ell_1 = \frac{-\lambda}{2\pi} \tan^{-1}\left(\frac{1}{b_1}\right) = 0.068\lambda \text{ or } \ell_1 = 0.164\lambda$$

$$\ell_2 = \frac{-\lambda}{2\pi} \tan^{-1}\left(\frac{1}{b_2}\right) = 0.051\lambda \text{ or } \ell_2 = 0.375\lambda$$

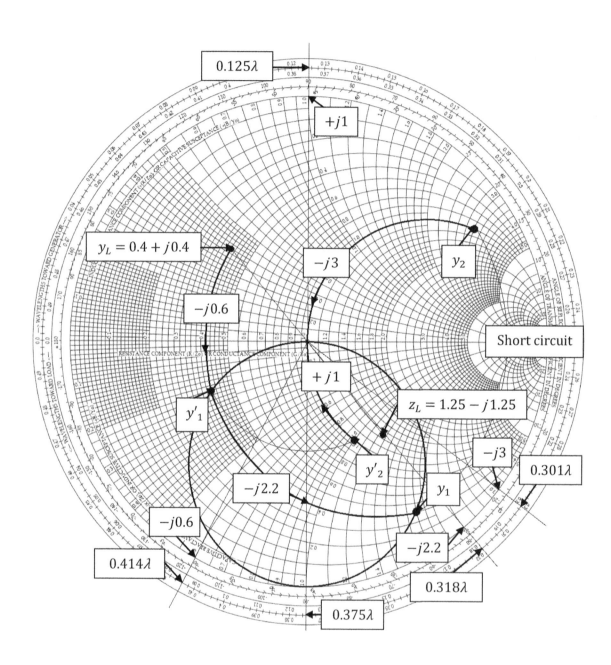

2. For $Z_L = 25 - j75 \ \Omega$, $Z_0 = 50 \ \Omega$:

$y_L = 0.2 + j0.6$

The susceptance of the first stub:

$b_1 = -2.2$ or $b'_1 = -1$

The susceptance of the second stub:

$b_2 = -4$ or $b'_2 = +2$

The lengths of the short-circuited stubs are:

$$\ell_1 = 0.068\lambda, \ \ell_2 = 0.039\lambda \quad \text{or} \quad \ell_1 = 0.125\lambda, \ \ell_2 = 0.426\lambda$$

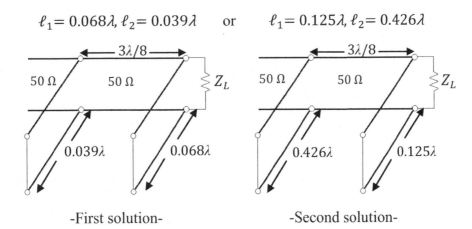

-First solution-   -Second solution-

Note: Analytically:

$t = tan(\beta d) = -1$

$b_1 = -b_L + \dfrac{1 \pm \sqrt{(1+t^2)g_L - g_L^2 t^2}}{t} \ \rightarrow b_1 = -2.2$ or $b_1 = -1$

$b_2 = \dfrac{\pm \sqrt{(1+t^2)g_L - g_L^2 t^2} + g_L}{g_L t} \ \rightarrow b_2 = -4$ or $b_2 = +2$

$\ell_1 = \dfrac{-\lambda}{2\pi} tan^{-1}(\dfrac{1}{b_1}) = 0.068\lambda$ or $\ell_1 = 0.125\lambda$

$\ell_2 = \dfrac{-\lambda}{2\pi} tan^{-1}(\dfrac{1}{b_2}) = 0.039\lambda$ or $\ell_2 = 0.426\lambda$

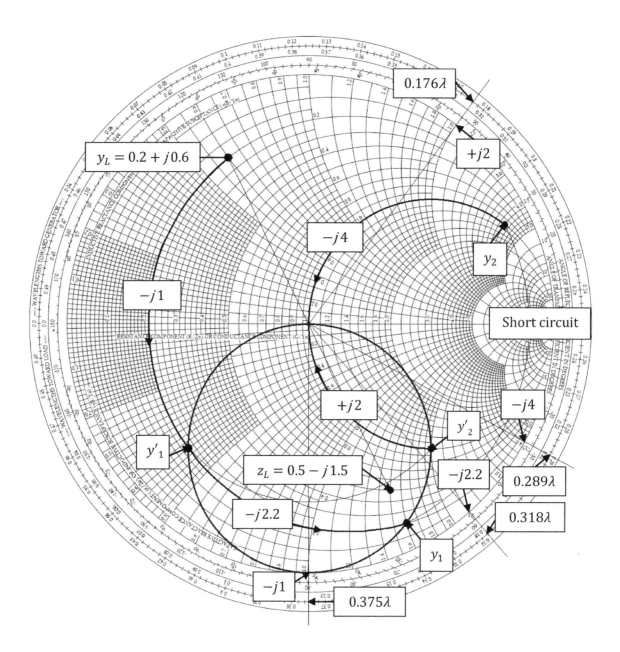

Problem 8. Using a length ℓ of a lossless transmission line of characteristic impedance Z_1, match a load $Z_L = 150 + j70\ \Omega$ to a 50 Ω line.

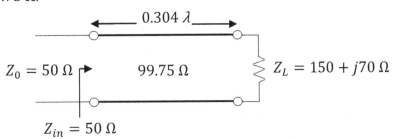

SOLUTION

To match Z_L to the 50 Ω line, Z_1 and ℓ must be determined so that $Z_{in} = Z_0 = 50\ \Omega$.

$$Z_{in} = Z_1 \frac{Z_L + j Z_1 \tan \beta \ell}{Z_1 + j Z_L \tan \beta \ell} = 50 = Z_1 \frac{(150 + j70) + j Z_1 t}{Z_1 + j(150 + j70)t} \quad \text{with} \quad t = \tan(\beta \ell)$$

$$(50 Z_1 - 3500t) + j7500t = 150 Z_1 + j(70 + Z_1 t) Z_1$$

By equating the real and imaginary parts, two equations for Z_1 and ℓ are found:

Re: $\quad 50 Z_1 - 3500t = 150 Z_1 \quad \rightarrow \quad Z_1 = -35t$

Im: $\quad 7500t = Z_1 (70 + Z_1 t)$

$$7500t = (-35t)(70 - 35t^2) \quad \rightarrow \quad t = \pm \sqrt{8.12} = \pm 2.85 \text{ (for } Z_1 > 0, \text{ use } -2.85)$$

Then, $\ell = \tan^{-1}(-2.85) \times \frac{\lambda}{2\pi} = 0.304\lambda \quad (\lambda/2 \text{ is added to get } \ell > 0)$.

The characteristic impedance is:

$Z_1 = -35\,(-2.85) = 99.75\ \Omega.$

Note: This method cannot be used to match all load impedances.

PART FOUR

MICROWAVE AMPLIFIERS

Problem 1. Consider two transistors with the following scattering parameters:

	S_{11}	S_{12}	S_{21}	S_{22}
Q_1	$0.78\angle - 90°$	$0.32\angle 69°$	$5.3\angle 80°$	$0.7\angle - 37°$
Q_2	$0.35\angle - 178°$	$0.07\angle 60°$	$4.15\angle 69°$	$0.45\angle - 32°$

Determine the stability of each device by using the $K - \Delta$ test and plot the stability circles if a device is potentially unstable.

SOLUTION

---

1. Q_1:

$$\Delta = S_{11}S_{22} - S_{12}S_{21} = 1.727\angle - 49°$$

$$K = \frac{1 - |S_{11}|^2 - |S_{22}|^2 + |\Delta|^2}{2\,|S_{12}S_{21}|} = 0.85$$

Since $|\Delta| > 1$ and $K < 1$, the unconditional stability criteria are not satisfied and the device is potentially unstable.

The centers and radii of the stability circles are:

$$C_L = \frac{(S_{22} - \Delta S_{11}^*)^*}{|S_{22}|^2 - |\Delta|^2} = 0.55\angle - 70°$$

$$R_L = \left| \frac{S_{12}S_{21}}{|S_{22}|^2 - |\Delta|^2} \right| = 0.68$$

$$C_S = \frac{(S_{11} - \Delta S_{22}^*)^*}{|S_{11}|^2 - |\Delta|^2} = 0.54\angle - 24°$$

$$R_S = \left| \frac{S_{12}S_{21}}{|S_{11}|^2 - |\Delta|^2} \right| = 0.71$$

The input and output stability circles are shown in the following figure:

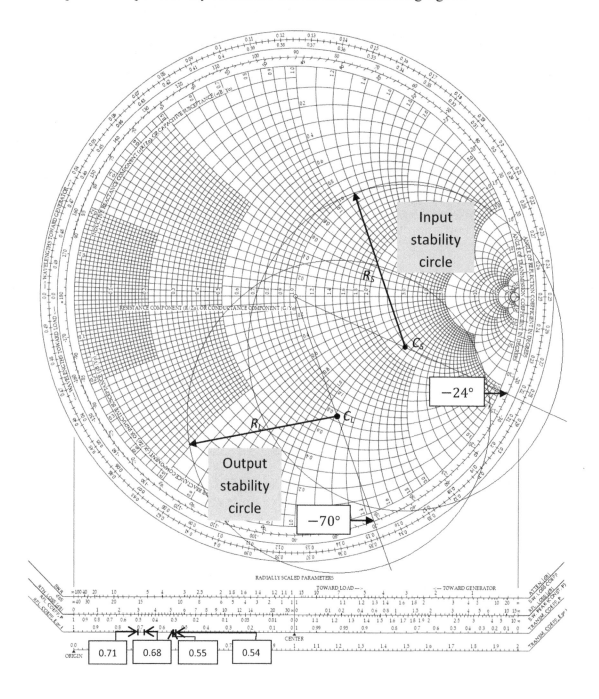

2. Q_2:

$\Delta = 0.154\angle - 72°$

The centers and radii of the stability circles are:

$C_L = 2.75\angle 36°$

$R_L = 1.62$

$C_S = 4.09\angle - 175°$

$R_S = 2.94$

Rollet's condition:

$K = 1.20$

Since $|\Delta| < 1$ and $K > 1$, the transistor is unconditionally stable.

Problem 2. Design an amplifier for maximum gain at 2.0 GHz with a GaAs FET that has the following S parameters ($Z_0 = 50\ \Omega$):

S_{11}	S_{12}	S_{21}	S_{22}
$0.75\angle -130°$	$0.03\angle 60°$	$2.6\angle 70°$	$0.6\angle -55°$

Open-circuited shunt stubs are to be considered in the design of the matching sections.

SOLUTION

First, check the transistor's stability:

$$\Delta = S_{11}S_{22} - S_{12}S_{21} = 0.399\angle -177°$$

$$K = \frac{1 - |S_{11}|^2 - |S_{22}|^2 + |\Delta|^2}{2\,|S_{12}S_{21}|} = 1.52$$

Since $|\Delta| < 1$ and $K > 1$, the transistor is unconditionally stable at 2.0 GHz.

For maximum gain, the transistor must be conjugately matched ($\Gamma_S = \Gamma_{in}{}^*$ and $\Gamma_L = \Gamma_{out}{}^*$):

Γ_S:

$$\Gamma_S = \frac{B_1 \pm \sqrt{B_1{}^2 - 4|C_1|^2}}{2\,C_1} = 0.842\angle 134°$$

With:

$$B_1 = 1 + |S_{11}|^2 - |S_{22}|^2 - |\Delta|^2$$

$$C_1 = S_{11} - \Delta S_{22}{}^*$$

Γ_L:

$$\Gamma_L = \frac{B_2 \pm \sqrt{B_2{}^2 - 4|C_2|^2}}{2\,C_2} = 0.751\angle 63°$$

With:

$$B_2 = 1 + |S_{22}|^2 - |S_{11}|^2 - |\Delta|^2$$

$$C_2 = S_{22} - \Delta S_{11}{}^*$$

Then, the gains are:

$$G_S = \frac{1}{1-|\Gamma_S|^2} = 3.436$$

$$G_0 = |S_{21}|^2 = 6.76$$

$$G_L = \frac{1-|\Gamma_L|^2}{|1-S_{22}\Gamma_L|^2} = 1.4037$$

The overall transducer gain is:

$$G_T = 3.436 \times 6.76 \times 1.4037 = 32.6 = 15.1 \text{ dB}$$

The matching networks are determined using the Smith chart as follows:

Step 1: Locate and plot $\Gamma_S{}^* = 0.842\angle - 134°$ on the Smith chart.

Step 2: Convert to the normalized admittance $y_S{}^*$ and work forward (toward the generator on the Smith chart) to find that a line of length 0.109λ will bring us to the $1+jb$ circle.

Step 3: Find that the required stub admittance is $+j\,3.14$, for an open-circuited stub length of 0.201λ.

Repeat this procedure with $\Gamma_L{}^*$ and find a line length of 0.105λ and a stub length of 0.316λ for the output matching circuit.

The final amplifier circuit is:

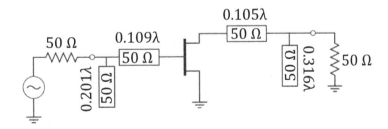

111

Γ_S^* :

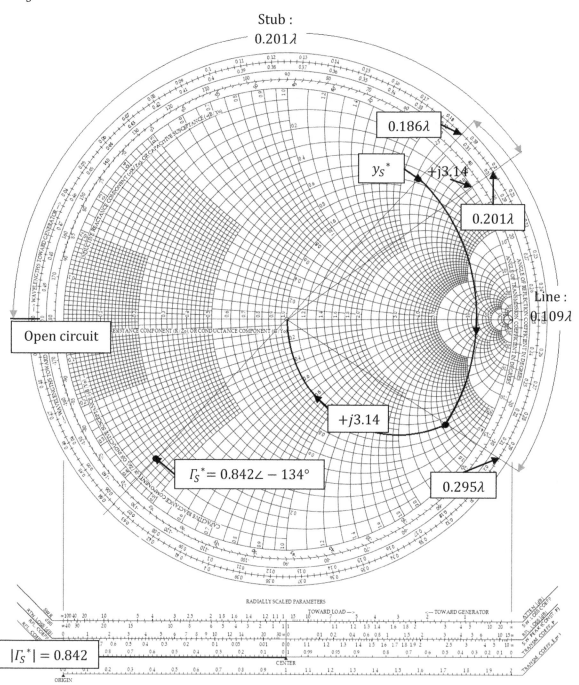

Γ_L^* :

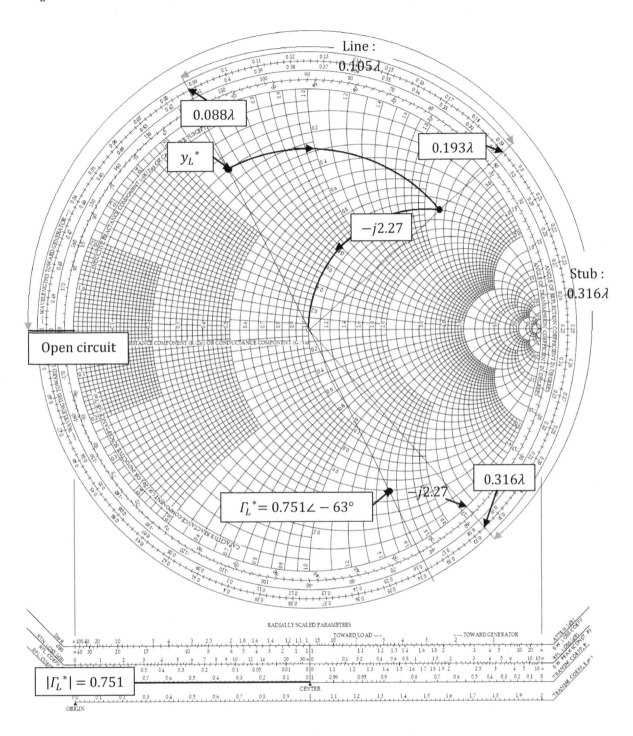

Note: The matching networks can also be determined starting from Γ_S and Γ_L. In this case, work backward (toward the load on the Smith chart) as follows:

Γ_S :

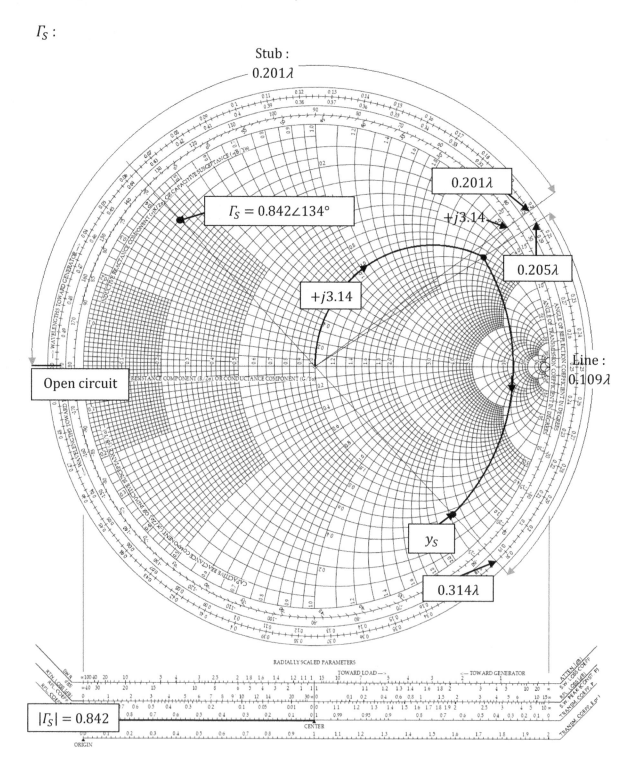

Γ_L :

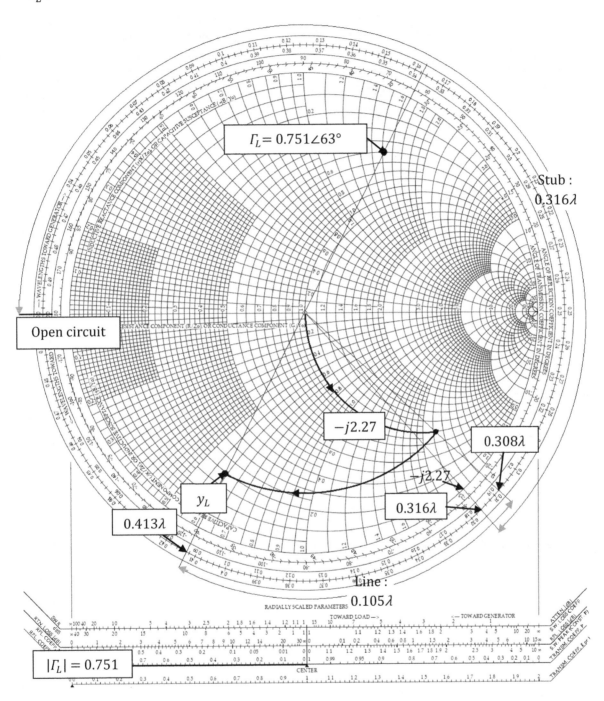

$\Gamma_L = 0.751 \angle 63°$

Stub :
0.316λ

Open circuit

$-j2.27$

0.308λ

$-j2.27$

y_L

0.316λ

0.413λ

Line :
0.105λ

RADIALLY SCALED PARAMETERS

$|\Gamma_L| = 0.751$

Problem 3. Repeat the previous problem for the following transistor S parameters (50 Ω):

S_{11}	S_{12}	S_{21}	S_{22}
$0.69\angle -135°$	$0.04\angle 61°$	$2.5\angle 65°$	$0.64\angle -60°$

SOLUTION

$$\Delta = S_{11}S_{22} - S_{12}S_{21} = 0.369\angle 175°$$

$$K = \frac{1-|S_{11}|^2-|S_{22}|^2+|\Delta|^2}{2\,|S_{12}S_{21}|} = 1.25$$

Since $|\Delta| < 1$ and $K > 1$, the transistor is unconditionally stable.

For maximum gain, the transistor must be conjugately matched (see previous problem):

$$\Gamma_S = 0.849\angle 140°$$

$$\Gamma_L = 0.825\angle 66°$$

The gains are:

$$G_S = 3.58$$

$$G_0 = 6.25$$

$$G_L = 1.39$$

The overall transducer gain is:

$$G_T = 31.1 = 14.9 \text{ dB}$$

The final amplifier circuit is:

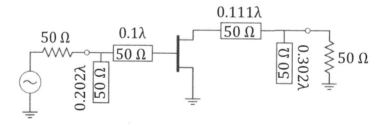

Γ_S^* :

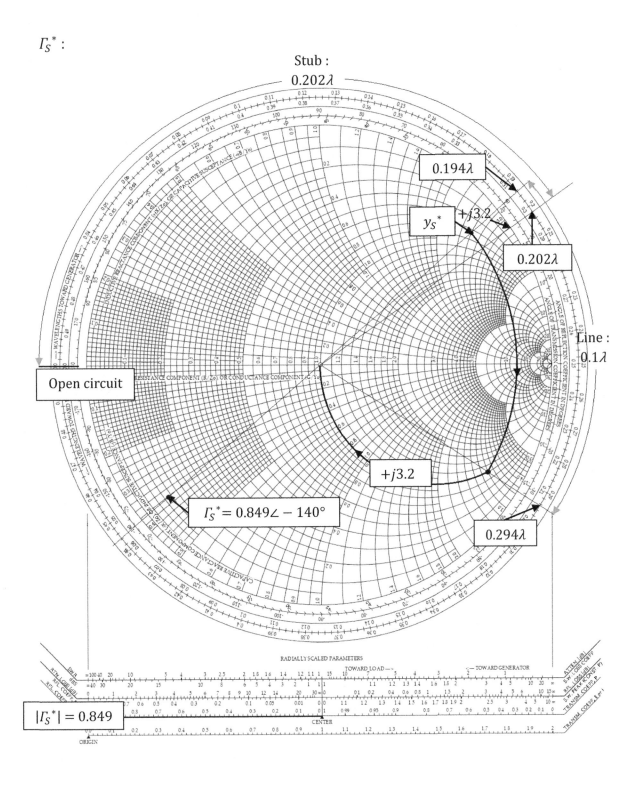

Stub :
0.202λ

0.194λ

y_S^* $+j3.2$

0.202λ

Line :
0.1λ

Open circuit

$+j3.2$

$\Gamma_S^* = 0.849\angle -140°$

0.294λ

$|\Gamma_S^*| = 0.849$

Γ_L^* :

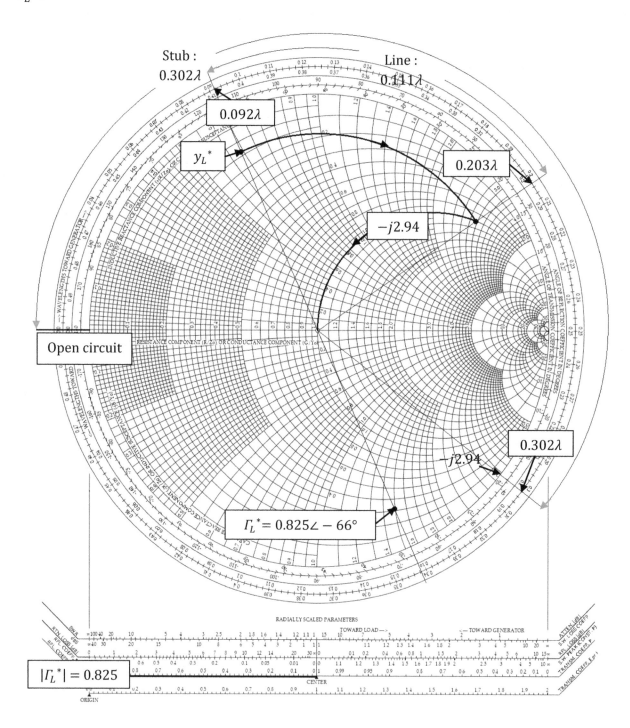

Problem 4. A microwave amplifier is to be designed for maximum G_{TU} using a transistor with the following S parameters at 1.0 GHz ($Z_0 = 50\ \Omega$):

S_{11}	S_{12}	S_{21}	S_{22}
$0.55\angle 142°$	0	$4\angle 47°$	$0.6\angle -92°$

1. Determine $G_{TU,\text{max}}$.
2. Design matching sections using open-circuited shunt stubs.
3. Draw the constant gain circle for $G_L = 1$ dB.

SOLUTION

Since $|\Delta| < 1$ and $K = \infty$, the transistor is unconditionally stable at 1.0 GHz. Since the transistor is unilateral:

1.

The maximum gains of the matching sections are:

$$G_{S,max} = \frac{1}{1-|S_{11}|^2} = 1.434 = 1.57\ \text{dB}$$

$$G_{L,max} = \frac{1}{1-|S_{22}|^2} = 1.563 = 1.94\ \text{dB}$$

The gain of the mismatched transistor is:

$$G_0 = |S_{21}|^2 = 16 = 12.04\ \text{dB}$$

Then, the maximum unilateral transducer power gain is:

$$G_{TU,max} = 1.57 + 1.94 + 12.04 = 15.55\ \text{dB}$$

2.

$$\Gamma_S = S_{11}{}^* = 0.55\angle -142°$$

$$\Gamma_L = S_{22}{}^* = 0.6\angle 92°$$

The matching networks are determined using the Smith chart as follows:

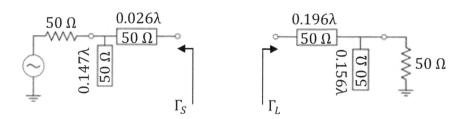

Γ_S^* :

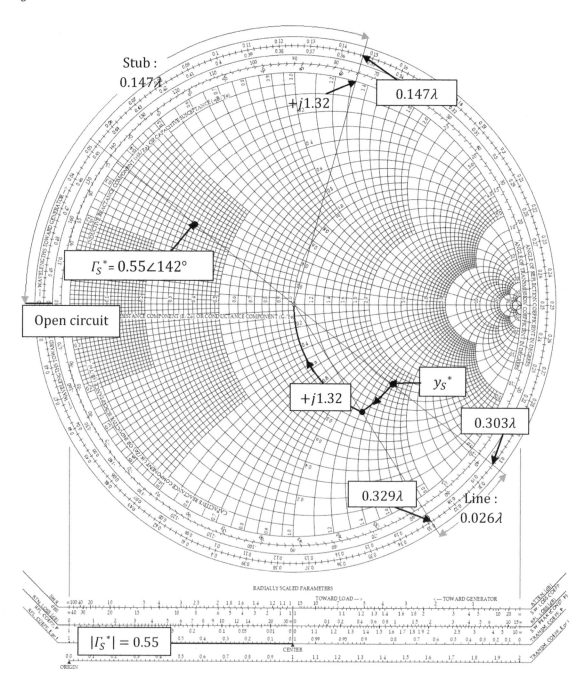

$\Gamma_L{}^*$:

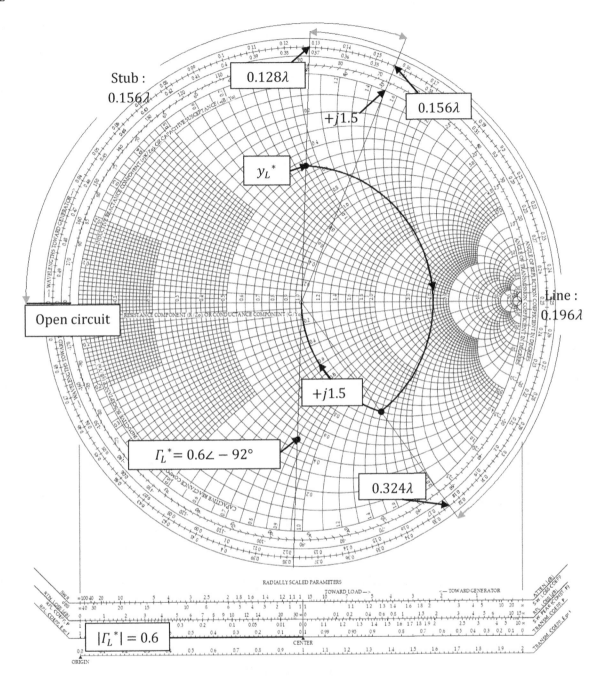

3.

$$G_L = 1 \text{ dB} = 10^{1/10} = 1.259$$

$$g_L = \frac{G_L}{G_{L,max}} = \frac{1.259}{1.563} = 0.806$$

$$C_L = \frac{g_L \, S_{22}{}^*}{1 - (1 - g_L) \, |S_{22}|^2} = 0.52\angle 92°$$

$$R_L = \frac{\sqrt{1 - g_L} \, (1 - |S_{22}|^2)}{1 - (1 - g_L) \, |S_{22}|^2} = 0.303$$

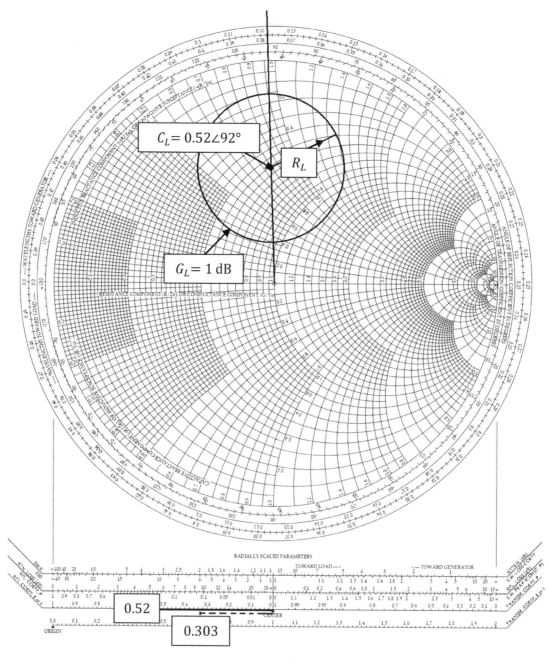

Problem 5. Repeat the previous problem for the following transistor S parameters (50 Ω):

S_{11}	S_{12}	S_{21}	S_{22}
$0.52\angle 164°$	0	$5\angle 43°$	$0.7\angle -98°$

SOLUTION

1.

$$G_{S,max} = \frac{1}{1-|S_{11}|^2} = 1.371 = 1.37 \text{ dB}$$

$$G_{L,max} = \frac{1}{1-|S_{22}|^2} = 1.961 = 2.92 \text{ dB}$$

$$G_0 = |S_{21}|^2 = 25 = 13.98 \text{ dB}$$

The maximum unilateral transducer power gain is:

$$G_{TU,max} = 1.37 + 2.92 + 13.98 = 18.27 \text{ dB}$$

2.

$$\Gamma_S = S_{11}^* = 0.52\angle -164°$$

$$\Gamma_L = S_{22}^* = 0.7\angle 98°$$

The matching networks are determined using the Smith chart as follows:

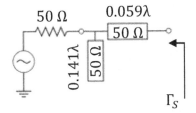
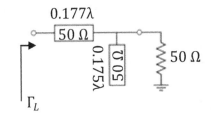

Γ_S^* :

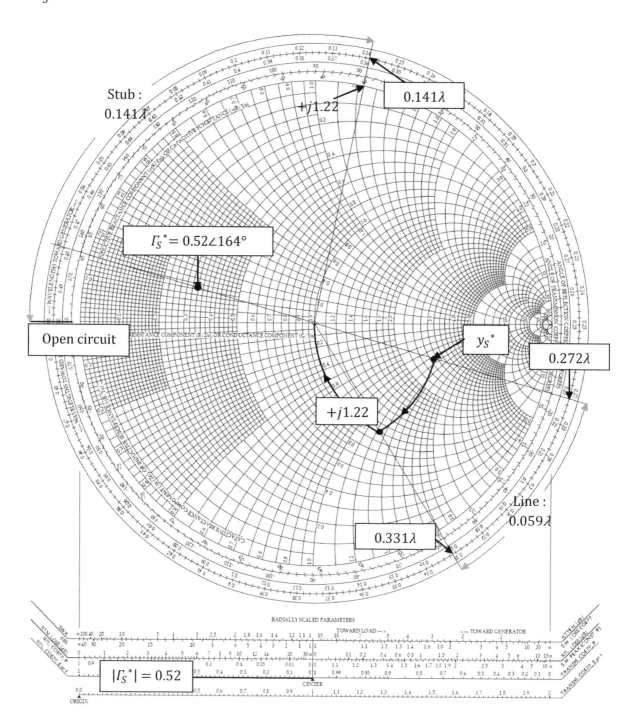

Stub :
0.141λ

0.141λ

$+j1.22$

$\Gamma_S^* = 0.52\angle164°$

Open circuit

y_S^*

0.272λ

$+j1.22$

Line :
0.059λ

0.331λ

RADIALLY SCALED PARAMETERS

$|\Gamma_S^*| = 0.52$

Γ_L^* :

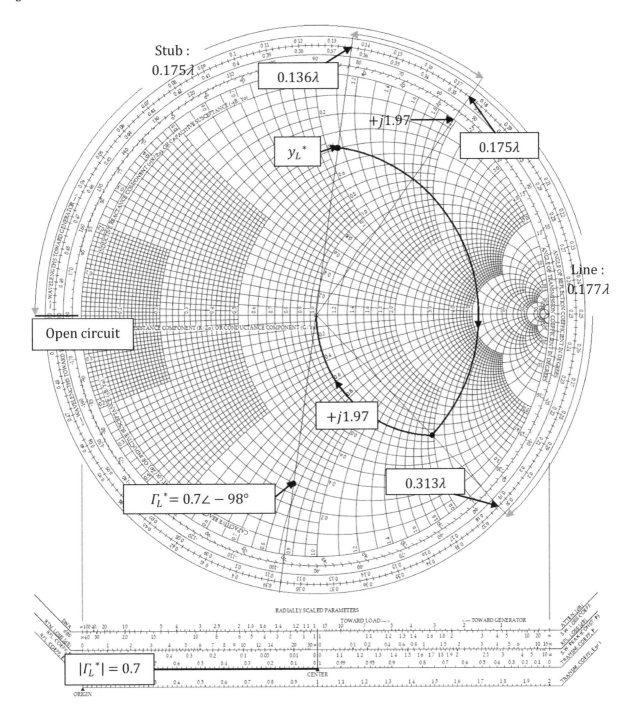

3.

$$g_L = \frac{G_L}{G_{L,max}} = \frac{1.259}{1.961} = 0.642$$

$$C_L = \frac{g_L\, S_{22}^{\;*}}{1-(1-g_L)\,|S_{22}|^2} = 0.545\angle 98°$$

$$R_L = \frac{\sqrt{1-g_L}\,(1-|S_{22}|^2)}{1-(1-g_L)\,|S_{22}|^2} = 0.37$$

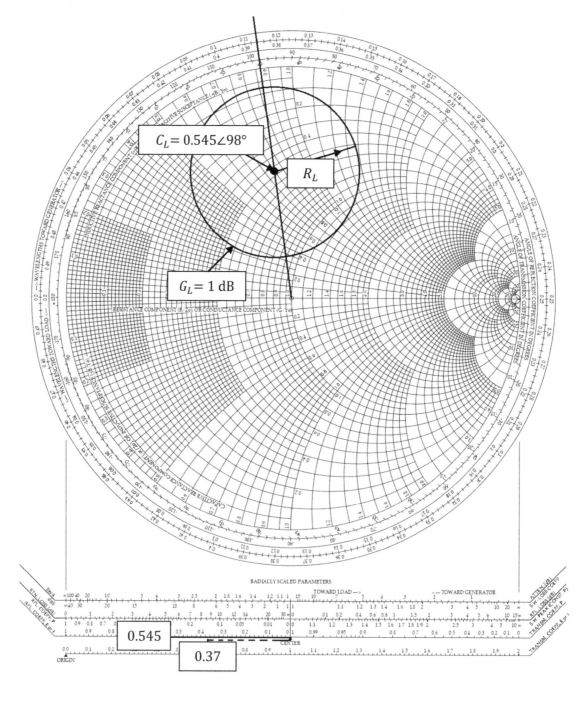

Problem 6. Consider a silicon bipolar junction transistor with the following S parameters at 1.5 GHz ($Z_0 = 50\ \Omega$):

S_{11}	S_{12}	S_{21}	S_{22}
$0.34\angle -147°$	$0.1\angle 51°$	$3.4\angle 83°$	$0.42\angle -47°$

For source impedance $Z_S = 25\ \Omega$ and load impedance $Z_L = 100\ \Omega$, calculate:

1. The power gain.
2. The available power gain.
3. The transducer power gain.

SOLUTION

The reflection coefficients at the source and load are:

$$\Gamma_S = \frac{Z_S - Z_0}{Z_S + Z_0} = \frac{25 - 50}{25 + 50} = -0.333$$

$$\Gamma_L = \frac{Z_L - Z_0}{Z_L + Z_0} = \frac{100 - 50}{100 + 50} = 0.333$$

The reflection coefficients seen looking at the input and output of the terminated network are:

$$\Gamma_{in} = S_{11} + \frac{S_{12}S_{21}\Gamma_L}{1 - S_{22}\Gamma_L} = 0.371\angle -167°$$

$$\Gamma_{out} = S_{22} + \frac{S_{12}S_{21}\Gamma_S}{1 - S_{11}\Gamma_S} = 0.545\angle -46°$$

1. The power gain is:

$$G = \frac{|S_{21}|^2(1 - |\Gamma_L|^2)}{(1 - |\Gamma_{in}|^2)|1 - S_{22}\Gamma_L|^2} = 14.4$$

2. The available power gain is:

$$G_A = \frac{|S_{21}|^2(1 - |\Gamma_S|^2)}{|1 - S_{11}\Gamma_S|^2(1 - |\Gamma_{out}|^2)} = 17.8$$

3. The transducer power gain:

$$G_T = \frac{|S_{21}|^2(1 - |\Gamma_S|^2)(1 - |\Gamma_L|^2)}{|1 - \Gamma_S\Gamma_{in}|^2|1 - S_{22}\Gamma_L|^2} = 14.2$$

127

Problem 7. Repeat the previous problem for the following transistor S parameters (50 Ω):

S_{11}	S_{12}	S_{21}	S_{22}
$0.29\angle - 140°$	$0.09\angle 57°$	$3.2\angle 80°$	$0.4\angle - 42°$

SOLUTION

$$\Gamma_S = -0.333$$

$$\Gamma_L = 0.333$$

$$\Gamma_{in} = 0.311\angle - 160°$$

$$\Gamma_{out} = 0.503\angle - 41°$$

1. The power gain: $G = 12.3$

2. The available power gain: $G_A = 14.2$

3. The transducer power gain: $G_T = 12.1$

Part five
Power dividers and directional couplers

Problem 1. Consider a directional coupler with the following scattering matrix:

$$[S] = \begin{bmatrix} 0.056\angle40° & 0.933\angle90° & 0.1\angle180° & 0.0017\angle90° \\ 0.933\angle90° & 0.056\angle40° & 0.0017\angle90° & 0.1\angle180° \\ 0.1\angle180° & 0.0017\angle90° & 0.056\angle40° & 0.933\angle90° \\ 0.0017\angle90° & 0.1\angle180° & 0.933\angle90° & 0.056\angle40° \end{bmatrix}$$

Calculate the return loss, directivity, coupling, isolation and insertion loss at the input port assuming that the other ports are terminated in matched loads.

SOLUTION

The return loss: $RL = -20\ log|\Gamma| = -20\ log|S_{11}| = -20\ log|0.056| = 25$ dB

Directivity: $D = 10\ log\frac{P_3}{P_4} = 20\ log\left|\frac{S_{13}}{S_{14}}\right| = 20\ log(\frac{0.1}{0.0017}) = 35$ dB

Coupling: $C = 10\ log\frac{P_1}{P_3} = -20\ log|S_{13}| = -20\ log(0.1) = 20$ dB

Isolation: $I = 10\ log\frac{P_1}{P_4} = -20\ log|S_{14}| = -20\ log(0.0017) = 55$ dB

Insertion loss: $L = 10\ log\frac{P_1}{P_2} = -20\ log|S_{12}| = -20\ log(0.933) = 0.6$ dB

Problem 2. By combining two identical 90° couplers, the outputs can be identical to those of a single 3 dB hybrid as shown in the following figure:

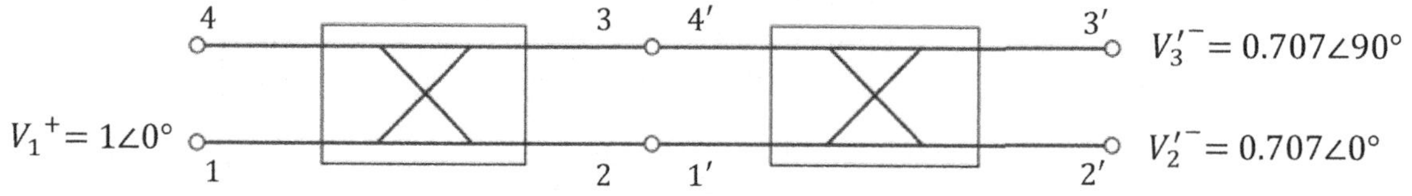

Find the coupling C that gives these outputs (at ports 2' and 3') relative to port 1.

SOLUTION

The scattering matrix in the case of a symmetric coupler ($\theta = \phi = \pi/2$) has the following form:

$$[S] = \begin{bmatrix} 0 & \alpha & j\beta & 0 \\ \alpha & 0 & 0 & j\beta \\ j\beta & 0 & 0 & \alpha \\ 0 & j\beta & \alpha & 0 \end{bmatrix}$$

From the latter:

$$V_3^- = j\beta \, V_1^+$$

$$V_2^- = \alpha \, V_1^+$$

The outputs of the second coupler (at ports 2' and 3') are:

$$V_3'^- = j\beta \, V_1'^+ + \alpha \, V_4'^+ = j\beta \, V_2^- + \alpha \, V_3^- = 0.707\angle 90°$$

$$V_2'^- = \alpha \, V_1'^+ + j\beta \, V_4'^+ = \alpha \, V_2^- + j\beta \, V_3^- = (\alpha^2 - \beta^2) \, V_1^+ = 0.707\angle 0°$$

From the output at port 2':

$$\alpha^2 - \beta^2 = \frac{0.707\angle 0°}{1\angle 0°} = 0.707$$

$\alpha^2 + \beta^2 = 1$, then:

$$\alpha^2 = 0.8535 \qquad \rightarrow \qquad \alpha = 0.924$$

$$\beta^2 = 0.1465 \qquad \rightarrow \qquad \beta = 0.383$$

$$C = -20 \, log \, \beta = -20 \, log \, 0.383 = 8.34 \text{ dB}$$

Problem 3. A lossless T-junction divider with 75 Ω source impedance is to be designed. The desired power split is 3:1.

1. Design quarter-wave matching transformers to convert the impedances of the output lines to 75 Ω.
2. Find the magnitude of the scattering parameters for this circuit using a 75 Ω characteristic impedance.

SOLUTION

1.

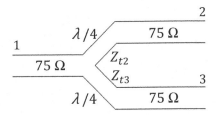

$$P_1 = \frac{1}{2}\frac{V_0^2}{Z_0}$$

$$P_2 = \frac{1}{2}\frac{V_0^2}{Z_2} = \frac{3}{4}P_1 = \frac{1}{2}V_0^2\left(\frac{3}{4Z_0}\right)$$

$$P_3 = \frac{1}{2}\frac{V_0^2}{Z_3} = \frac{1}{4}P_1 = \frac{1}{2}V_0^2\left(\frac{1}{4Z_0}\right)$$

Then,

$$Z_2 = \frac{4Z_0}{3} = 100\ \Omega$$

$$Z_3 = 4Z_0 = 300\ \Omega$$

The impedances of the $\lambda/4$ matching transformers are:

$$Z_{t2} = \sqrt{Z_0 Z_2} = \sqrt{75(100)} = 86.6\ \Omega$$

$$Z_{t3} = \sqrt{Z_0 Z_3} = \sqrt{75(300)} = 150\ \Omega$$

2. The *S*-Parameters are:

$$S_{11} = \frac{75-75}{75+75} = 0$$

$$S_{22} = \frac{75||300-100}{75||300+100} = \frac{60-100}{60+100} = -0.25$$

$$S_{33} = \frac{75||100-300}{75||100+300} = \frac{42.86-300}{42.86+300} = -0.75$$

$$S_{21} = S_{12} = \sqrt{\frac{P_2}{P_1}}\, e^{-j\theta} = \sqrt{\frac{3}{4}} \angle -90° = 0.866 \angle -90°$$

$$S_{31} = S_{13} = \sqrt{\frac{P_3}{P_1}}\, e^{-j\theta} = \sqrt{\frac{1}{4}} \angle -90° = 0.5 \angle -90°$$

Since the T-junction divider is lossless:

$$|S_{21}|^2 + |S_{22}|^2 + |S_{23}|^2 = 1$$

Then,

$$S_{23} = S_{32} = \sqrt{1 - (0.25)^2 - (0.866)^2}\, e^{-2j\theta} = 0.433 \angle -180°$$

Problem 4. Repeat the previous problem for a 2:1 ratio power split.

SOLUTION

1. The impedances of the $\lambda/4$ matching transformers:

$$P_1 = \frac{1}{2}\frac{V_0^2}{Z_0}$$

$$P_2 = \frac{1}{2}\frac{V_0^2}{Z_2} = \frac{2}{3}P_1 = \frac{1}{2}V_0^2\left(\frac{2}{3Z_0}\right) \qquad \rightarrow \qquad Z_2 = \frac{3Z_0}{2} = 112.5\ \Omega$$

$$P_3 = \frac{1}{2}\frac{V_0^2}{Z_3} = \frac{1}{3}P_1 = \frac{1}{2}V_0^2\left(\frac{1}{3Z_0}\right) \qquad \rightarrow \qquad Z_3 = 3Z_0 = 225\ \Omega$$

$$Z_{t2} = 91.86\ \Omega$$

$$Z_{t3} = 129.9\ \Omega$$

2. The S-Parameters are:

$$S_{11} = 0$$

$$S_{22} = -0.333$$

$$S_{33} = -0.667$$

$$S_{21} = S_{12} = 0.816\angle -90°$$

$$S_{31} = S_{13} = 0.577\angle -90°$$

$$S_{23} = S_{32} = 0.472\angle -180°$$

Problem 5. Design a Wilkinson power divider with a power division ratio of $P_3/P_2 = 2/3$ and a source impedance of 50 Ω.

SOLUTION

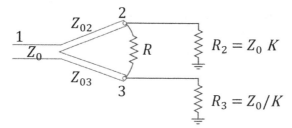

$$K^2 = \frac{P_3}{P_2} = \frac{2}{3} \qquad \rightarrow \qquad K = 0.816$$

$$Z_{03} = Z_0 \sqrt{\frac{1+K^2}{K^3}} = 87.5 \ \Omega$$

$$Z_{02} = K^2 Z_{03} = 58.3 \ \Omega$$

$$R = Z_0 (K + 1/K) = 102.1 \ \Omega$$

The output impedances are:

$$R_2 = Z_0 K = 40.8 \ \Omega$$

$$R_3 = Z_0/K = 61.2 \ \Omega$$

Problem 6. Repeat the previous problem for a power division ratio of $P_3/P_2 = 1/3$ and a source impedance of 75 Ω.

SOLUTION

$$K^2 = \frac{P_3}{P_2} = \frac{1}{3} \qquad \rightarrow \qquad K = 0.577$$

$$Z_{03} = 197.4 \ \Omega$$

$$Z_{02} = 65.8 \ \Omega$$

$$R = 173.2 \ \Omega$$

The output impedances are:

$$R_2 = 43.3 \ \Omega$$

$$R_3 = 129.9 \ \Omega$$

Problem 7. Design an equal-split Wilkinson power divider for a 50 Ω system impedance.

The characteristic impedance of the quarter-wave transmission lines in the divider are:

$$Z = \sqrt{2}\, Z_0 = \sqrt{2}\, 50 = 70.7 \ \Omega$$

The shunt resistor value is:

$$R = 2\, Z_0 = 2 \times 50 = 100 \ \Omega$$

Problem 8. Consider a 180° hybrid where input signals V_1 and V_4 are applied to the sum port and difference port, respectively. Determine the output signals.

The scattering matrix of a 3 dB 180° hybrid is:

$$[S] = \frac{-j}{\sqrt{2}} \begin{bmatrix} 0 & 1 & 1 & 0 \\ 1 & 0 & 0 & -1 \\ 1 & 0 & 0 & 1 \\ 0 & -1 & 1 & 0 \end{bmatrix}$$

The 180° hybrid is matched ($V_1 = V_1^+$ and $V_4 = V_4^+$).

The output voltages are:

$$\begin{bmatrix} V_1^- \\ V_2^- \\ V_3^- \\ V_4^- \end{bmatrix} = \frac{-j}{\sqrt{2}} \begin{bmatrix} 0 & 1 & 1 & 0 \\ 1 & 0 & 0 & -1 \\ 1 & 0 & 0 & 1 \\ 0 & -1 & 1 & 0 \end{bmatrix} \begin{bmatrix} V_1 \\ 0 \\ 0 \\ V_4 \end{bmatrix}$$

$$\begin{bmatrix} V_1^- \\ V_2^- \\ V_3^- \\ V_4^- \end{bmatrix} = \frac{-j}{\sqrt{2}} \begin{bmatrix} 0 \\ V_1 - V_4 \\ V_1 + V_4 \\ 0 \end{bmatrix}$$

PART SIX
QUICK TEST

Problem 1

A 50 Ω transmission line of electrical length $\ell = 0.147\lambda$ is terminated with a complex load impedance $Z_L = 25 + j40\ \Omega$. Calculate:

1. The reflection coefficient at the load.
2. The input impedance of the line.
3. The reflection coefficient at the input of the line.
4. The SWR on the line.

Problem 2

An amplifier is to be designed for maximum gain at 2.0 GHz with a GaAs FET that has the following S parameters ($Z_0 = 50\ \Omega$):

S_{11}	S_{12}	S_{21}	S_{22}
$0.7\angle - 135°$	$0.03\angle 65°$	$2.6\angle 70°$	$0.6\angle - 50°$

1. Check the transistor's stability.
2. Design the matching sections using open-circuited shunt stubs.

Problem 3

Design a bandpass maximally flat lumped-element filter with the following specifications:

Order	Center frequency	Bandwidth	Impedance
$N = 3$	$f_0 = 1.5$ GHz	100 MHz	$Z_0 = 50\ \Omega$

Problem 4

1. Identify this component based on its scattering matrix ($Z_0 = 50\ \Omega$):

$$[S] = \frac{-j}{\sqrt{2}}\begin{bmatrix} 0 & 1 & 1 & 0 \\ 1 & 0 & 0 & -1 \\ 1 & 0 & 0 & 1 \\ 0 & -1 & 1 & 0 \end{bmatrix}$$

2. Are all ports completely matched to 50 Ω?
3. Is there an isolated port?
4. If 10 dBm is applied to port 3, determine the output signals at the other ports in dBm, watts and volts.

SOLUTION

Problem 1:

1) $\Gamma_L = 0.555 \angle 94°$

2) $Z_{in} = 155.9 - j51.6 \, \Omega$

3) $\Gamma_{in} = 0.555 \angle 348°$

4) $SWR = 3.49$

Problem 2:

$\Delta = 0.364\angle - 177°$

$K = 1.81$

Since $|\Delta| < 1$ and $K > 1$, the transistor is unconditionally stable at 2.0 GHz.

For maximum gain, the transistor must be conjugately matched ($\Gamma_S = \Gamma_{in}^{*}$ and $\Gamma_L = \Gamma_{out}^{*}$):

$\Gamma_S = 0.786 \angle 139°$

$\Gamma_L = 0.719 \angle 56°$

The final amplifier circuit is:

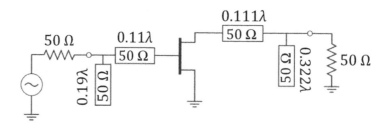

$\Gamma_S{}^*$:

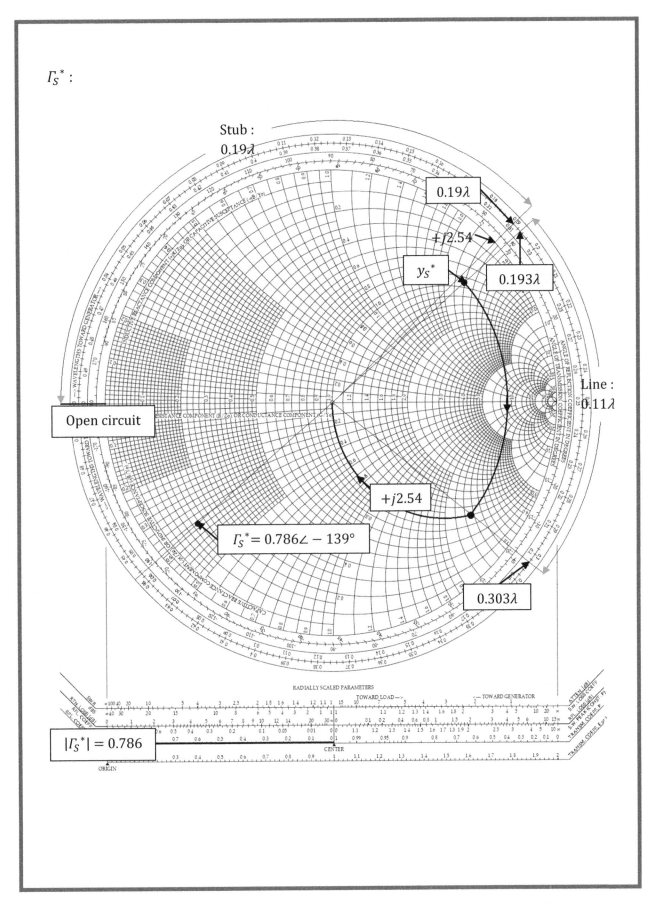

Stub :
0.19λ

0.19λ

+j2.54

$y_S{}^*$

0.193λ

Line :
0.11λ

Open circuit

+j2.54

$\Gamma_S{}^* = 0.786\angle - 139°$

0.303λ

RADIALLY SCALED PARAMETERS

$|\Gamma_S{}^*| = 0.786$

$\Gamma_L^*:$

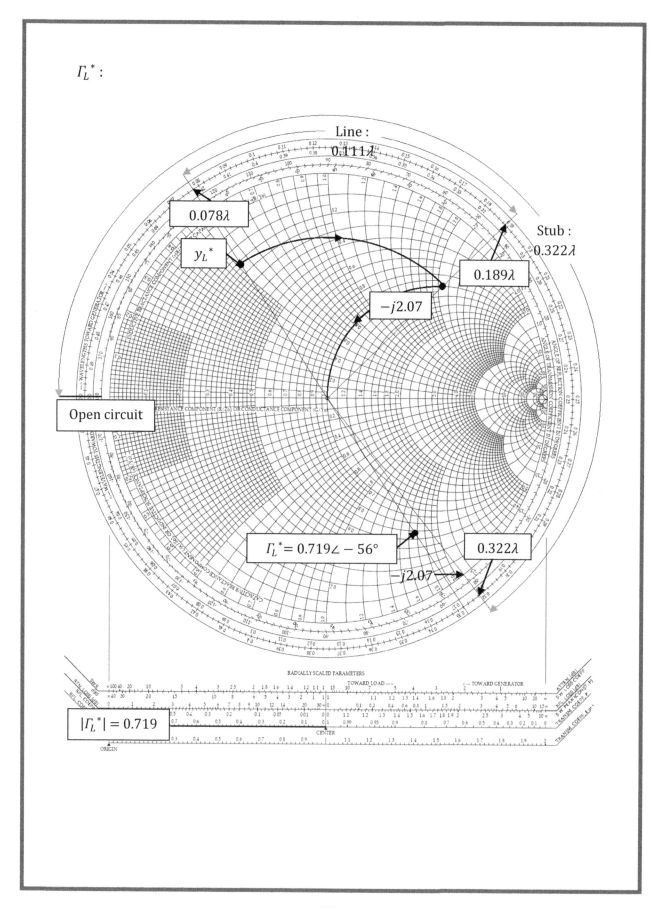

Line : 0.111λ

0.078λ

y_L^*

Stub : 0.322λ

0.189λ

$-j2.07$

Open circuit

$\Gamma_L^* = 0.719\angle -56°$

0.322λ

$-j2.07 \longrightarrow$

RADIALLY SCALED PARAMETERS

$|\Gamma_L^*| = 0.719$

Problem 3:

$g_1 = 1, \ g_2 = 2, \ g_3 = 1$

$\Delta = 0.06667$

$L_1 = 0.35 \ \text{nH}$

$C_1 = 31.83 \ \text{pF}$

$L_2 = 159.15 \ \text{nH}$

$C_2 = 0.07 \ \text{pF}$

$L_3 = 0.35 \ \text{nH}$

$C_3 = 31.83 \ \text{pF}$

The bandpass filter:

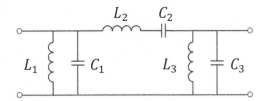

Problem 4:

1. This component is a 180° hybrid junction.
2. All ports are completely matched to $50 \ \Omega \rightarrow S_{11} = S_{22} = S_{33} = S_{44} = 0$
3. If a signal is applied to port 1, port 4 is isolated (the signal is evenly split into two in-phase components at ports 2 and 3). If the input is applied to port 4, port 1 is then isolated (the signal is equally split into two components with a 180° phase difference at ports 2 and 3).
4. $P_3 = \dfrac{1}{2} \dfrac{V_3{}^2}{Z_0}$

    From the scattering matrix:

    - $S_{13} = S_{43}$

    $\dfrac{P_1}{P_3} = \dfrac{P_4}{P_3} = \dfrac{1}{2}$

    $P_1 = P_4 = \dfrac{P_3}{2} = 5 \ \text{dBm} = 0.00316 \ \text{W}$

    $V = \sqrt{2 \times P \times Z_0} = 0.56 \ \text{V}$

    $\angle S_{13} = \angle S_{43} = -90°$

    - $S_{23} = 0 \rightarrow P_2 = 0 \ \text{W}$

    $\angle S_{23} = 0°$

NOTES

NOTES

NOTES

NOTES

NOTES

Made in the USA
Las Vegas, NV
14 August 2021